AS-Level

Biology

OCR

The Revision Guide

Editors:
Becky May, Kate Redmond.

Contributors:
Gloria Barnett, Claire Charlton, Martin Chester, Barbara Green, Anna-Fe Guy, Dominic Hall, Gemma Hallam, Kate Houghton, Stephen Phillips, Claire Reed, Katherine Reed, Adrian Schmit, Emma Singleton, Sharon Watson.

Proofreaders:
Ben Aldiss, Vanessa Aris, James Foster, Tom Trust.

Published by Coordination Group Publications Ltd.

ISBN: 1 84146 957 2
Groovy website: www.cgpbooks.co.uk
Jolly bits of clipart from CorelDRAW
Printed by Elanders Hindson, Newcastle upon Tyne.

Contents

Electron and Light Microscopy

You can't get away from microscopes in biology. I mean, here they are barging in on the very first page.
And you need to learn all about them, otherwise all these colourful splodges will remain meaningless for ever more.

Magnification is Size, Resolution is Detail

Different types of **microscopes** produce **images** with different properties.
The **magnification** and the **resolution** of the image are two of the most important properties.

1)   **Magnification** is how much bigger the microscope image is than the specimen.
It's calculated as: <u>length of drawing or photograph</u>
                      length of specimen object

2)   **Resolution** is how detailed the image is. More specifically, it's how well a microscope
distinguishes between two points that are close together. If a microscope lens
can't separate two objects, then increasing the magnification won't help.

Rob was rather pleased with the level of resolution he achieved using his new microscope.

There are two main types of microscope:

A micrometre (µm) is a thousandth of a millimetre. A nanometre (nm) is a thousandth of a micrometre. That's tiny.

Light microscopes have a **lower resolution** than electron microscopes.
Decreasing the wavelength of the light increases resolution, but even then a light
microscope can only distinguish points that are 0.2 micrometres (µm) apart.

Electron microscopes use **electrons** instead of light to form an image, and
focus them with an electromagnet. You can't see electrons, so the image has to
be formed on a fluorescent screen. Electrons have a much **shorter wavelength**
than light, and can resolve things down to 0.5 nanometres (0.0005 µm) — so
electron microscopes provide better resolution and **more detailed images**.

Light Microscopes Show Cell Structure

If you just want to see the **general structure of a cell**, then light microscopes are fine.
But even the best light microscopes can't see most of the organelles in the cell.
You can see the larger organelles, like the nucleus, but none of the internal details.

Organelles are the structures that you find inside a cell (see page 4).

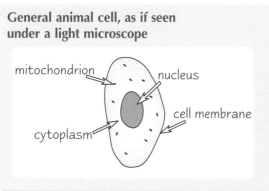

General animal cell, as if seen under a light microscope

mitochondrion   nucleus
cytoplasm   cell membrane

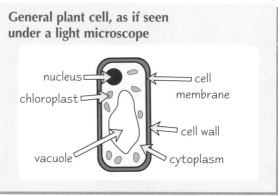

General plant cell, as if seen under a light microscope

nucleus   cell membrane
chloroplast   cell wall
vacuole   cytoplasm

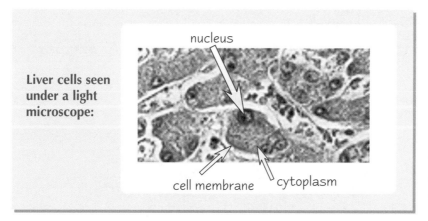

Liver cells seen under a light microscope:

nucleus
cell membrane   cytoplasm

To see these cells in more
detail, you'd need to use
an **electron microscope**.

Electron and Light Microscopy

Electron Microscopes Show Organelles

There's not much in a cell that an electron microscope can't see. You can see the **organelles** and the **internal structure** of most of them. Most of what's known about cell structure has been discovered by electron microscope studies. The diagrams below show what you can see in plant and animal cells under an electron microscope. Nice.

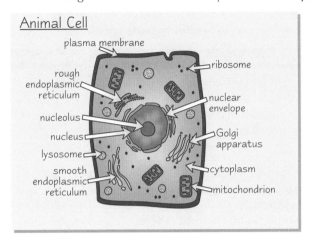

Animal Cell

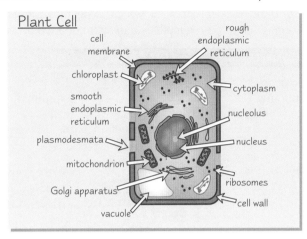

Plant Cell

Comparing Light and Electron Microscopy

LIGHT MICROSCOPE	ELECTRON MICROSCOPE
Resolution down to 0.2 μm	Resolution down to 0.5 nm
Maximum magnification is × 1500	Maximum magnification is × 250000
Living tissue can be examined	Processing kills living cells
Colours can be seen	Natural colours can't be seen
Mobile	Can't be moved around
Relatively cheap	Very expensive

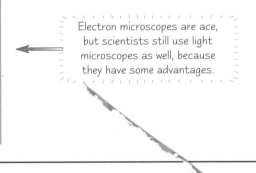

Electron microscopes are ace, but scientists still use light microscopes as well, because they have some advantages.

Practice Questions

Q1   How do you calculate magnification?

Q2   Why do electron microscopes have a better resolution than light microscopes?

Q3   Name two organelles seen under the light microscope in a plant cell that are not seen in animal cells.

Q4   Give three advantages of using a light microscope rather than an electron microscope.

Q5   Name two organelles that can be seen in an animal cell using light microscopes, and two that can't be seen.

Exam Question

Q1   Explain the advantages and disadvantages of using an electron microscope
      rather than a light microscope to study cells.

[6 marks]

Learn to use a microscope — everything will become clearer...

OK, so you've read the first topic and now it's the moment you've been waiting for — yep, it's time to get learning those facts. You need to know the differences between light and electron microscopes, and the advantages and disadvantages of using each one. And make sure you can recognise typical plant and animal cells. And brush your hair — you look like a mess.

Functions of Organelles

Organelles are all the tiny bits and bobs inside a cell that you can only see in detail with an electron microscope. It's cool to think that all these weird and wonderful things live inside our tiny cells.

Cells Contain **Organelles**

An organelle is a structure found inside a cell — each organelle has a specific function. Most organelles are surrounded by membranes, which sometimes causes confusion — don't make the mistake of thinking that a diagram of an organelle is a diagram of a whole cell. They're not cells — they're **parts of** cells, see.

ORGANELLE	DIAGRAM	DESCRIPTION	FUNCTION
Plasma membrane	plasma membrane / cytoplasm	The membrane found on the surface of **animal cells** and just inside the **cell wall** of **plant cells**. It's made mainly of **protein** and **lipids**.	**Regulates the movement** of substances into and out of the cell. It also has **receptor molecules** on it, which allow it to respond to chemicals like hormones.
Nucleus	nuclear envelope / nucleolus / nuclear pore / chromatin	A large organelle surrounded by a **nuclear membrane** or **envelope**, which contains many **pores**. The nucleus contains **chromatin** and often a structure called the **nucleolus**.	The **chromatin** contains the genetic material (DNA) which **controls the cell's activities**. The pores allow substances (e.g. RNA) to move between the nucleus and the cytoplasm. The **nucleolus** makes **RNA**.
Lysosome		A **round organelle** surrounded by a **membrane**, with no clear internal structure.	Contains **digestive enzymes**. These are kept separate from the cytoplasm by the surrounding membrane, but can be used to **digest invading cells** or to **destroy the cell** when it needs to be replaced.
Ribosome	small subunit / large subunit	A **very small organelle** either floating free in the cytoplasm or attached to the rough endoplasmic reticulum.	The **site** where **proteins** are made.
Rough Endoplasmic Reticulum (RER)	ribosome / fluid	A system of membranes enclosing a fluid-filled space. The surface is **covered with ribosomes**.	**Transports proteins** which have been made in the ribosomes.
Smooth Endoplasmic Reticulum		Similar to rough endoplasmic reticulum, but with no **ribosomes**.	**Transports lipids** around the cell.
Golgi Apparatus	vesicle	A group of smooth endoplasmic reticulum consisting of a series of **flattened sacs**. Vesicles are often seen at the edges of the sacs.	It **packages** substances that are produced by the cell, mainly proteins and glycoproteins. It also **makes lysosomes**.

Functions of Organelles

ORGANELLE	DIAGRAM	DESCRIPTION	FUNCTION
Mitochondrion	outer membrane, inner membrane, crista, matrix	They are usually oval. They have a **double membrane** — the inner one is folded to form structures called **cristae**. Inside is the **matrix**, which contains enzymes involved in respiration (but, sadly, no Keanu Reeves).	The **site of respiration**, where **ATP** is produced. They are found in large numbers in cells that are very active and require a lot of energy.
Chloroplast	stroma, two membranes, granum (plural = grana), lamella (plural = lamellae)	A small, **flattened** structure found in **plant cells**. It's surrounded by a **double membrane**, and also has membranes inside called **thylakoid membranes**. These membranes are stacked up in some parts of the chloroplast to form **grana**. Grana are linked together by lamellae — thin, flat pieces of thylakoid membrane.	The **site** where **photosynthesis** takes place. The light-dependent reaction of photosynthesis happens in the **grana**, and the light-independent reaction of photosynthesis happens in the **stroma**.
Centriole		Small, **hollow cylinders**, containing a ring of microtubules, seen in **animal cells** during cell division.	Involved with the **separation of chromosomes** during cell division
Cilia	side, cross-section	Small, **hair-like structures** found on the surface membrane of some **animal cells**. In cross section, they have an outer membrane and a ring of 9 pairs of **microtubules** inside, with a single pair of microtubules in the middle.	The microtubules allow the cilia to **move**. This movement is used by the cell to **move substances along the cell surface**.

Practice Questions

Q1 Which organelles have a double membrane?

Q2 Which organelle contains digestive enzymes?

Q3 Explain the differences between the rough and smooth endoplasmic reticulum.

Q4 What is the function of the nucleolus?

Exam questions

Q1 The presence and number of specific organelles can give an indication of a cell's function.
Give THREE examples of this, naming the organelles concerned and stating their function. [9 marks]

Q2 a) Identify these two organelles seen in an electron micrograph, from the descriptions given below.

　(i) A sausage-shaped organelle surrounded by a double membrane.
The inner membrane is folded and projects into the inner space,
which is filled with a grainy material.

　(ii) A collection of flattened membrane 'bags' arranged roughly parallel to one another.
Small circular structures are seen at the edges of these 'bags'. [2 marks]

　b) State the function of the two organelles that you have identified. [2 marks]

Organs and organelles — 'his and hers' biology terms...

Organelle is a very pretty-sounding name for all those blobs. But under a microscope some of them are actually quite fetching — well I think so anyway, but then my mate finds woodlice fetching, so there's no accounting for taste. Anyway, you need to know the names and functions of all the organelles and also what they look like under the microscope.

Cell Organisation

These pages tell you all about how nicely your cells are organised into tissues and organs. It's great to be organised, with lovely revision timetables in all different colours. But make sure you don't forget to do the actual learning.

There are **Two Types** of Cell — **Prokaryotic** and **Eukaryotic**

Prokaryotic cells are **simpler** than eukaryotic cells. Prokaryotes include **bacteria** and **blue-green algae**. **Eukaryotic** cells are more complex, and include all **animal and plant cells** (see diagrams of plant and animal cells on p.2-3.)

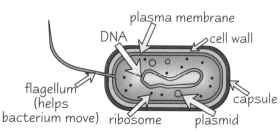

A Typical Prokaryotic Cell

PROKARYOTES	EUKARYOTES
Extremely small cells (0.5-3.0 μm diameter)	Larger cells (20-40 μm diameter)
No nucleus — DNA free in cytoplasm	Nucleus present
Cell wall made of a polysaccharide, but not cellulose or chitin	No cell wall (in animals), cellulose cell wall (in plants) or chitin cell wall (in fungi)
Few organelles, no mitochondria	Many organelles, mitochondria present
Small ribosomes	Larger ribosomes
Example: *E. coli* bacterium	Example: Human liver cell

Similar Cells are Organised into **Tissues**

A single-celled organism performs all its life functions in its one cell. **Multicellular organisms** (like us) are more complicated — different cells are adapted to do different jobs, so cells have to be **organised** into different groups. Similar cells are grouped together into **tissues**:

Squamous epithelium (plural = **epithelia**) is a **single layer** of **flat cells** lining a surface. It's pretty common in the body. The cells lining the **alveoli** are squamous epithelium cells.

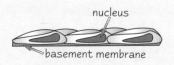

Epithelium means a tissue that forms a covering or a lining.

Ciliated epithelium has moving hair-like structures called **cilia** on it. It's found on surfaces where things need to be moved — in the trachea for instance, where the cilia waft mucus along.

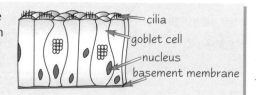

Tissues aren't always made up of just one type of cell. Some tissues include different types of cell working together.

To complicate things, you don't have epithelia in plants. The covering tissue is called the **epidermis** instead.

Xylem is a plant tissue with two jobs — it **transports water** around the plant, and it **supports** the plant. The cells are mostly **dead and hollow** with no end walls (so they are like tubes) and they have **thick walls** for strength.

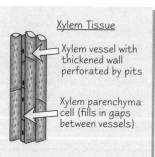

Xylem Tissue

Xylem vessel with thickened wall perforated by pits

Xylem parenchyma cell (fills in gaps between vessels)

Phloem tissue **carries sugars** around the plant. It's also arranged in tubes. Each cell has end walls with **holes** in them, so that sap can move easily through them. These end walls are called **sieve plates**.

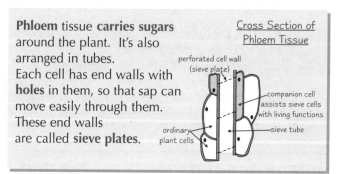

Cross Section of Phloem Tissue

perforated cell wall (sieve plate)

companion cell assists sieve cells with living functions

sieve tube

ordinary plant cells

Cell Organisation

Tissues are Organised into Organs

An **organ** is a **group of tissues** that work together to perform a particular function.

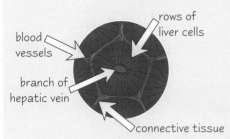

blood vessels
rows of liver cells
branch of hepatic vein
connective tissue

The **liver** is an example of an animal organ.
1) **Liver cells** are the main tissue.
2) There are blood vessels containing blood to provide food and oxygen for the liver cells. **Blood** is a tissue (yes, really).
3) **Connective tissue** holds the organ together.

Blood vessels aren't a tissue, though. They contain several tissues (epithelium, muscle etc.), so they're actually organs.

The **leaf** is an example of a plant organ. It's made up of the following tissues:

1) **Lower epidermis** — contains stomata (holes) to let carbon dioxide and oxygen in and out.
2) **Spongy mesophyll** — full of spaces to let gases circulate.
3) **Palisade mesophyll** — contains lots of chloroplasts. Most photosynthesis takes place here.
4) **Xylem** — carries water to the leaf.
5) **Phloem** — carries sugars away from leaf.
6) **Upper epidermis** — covered in a waterproof waxy cuticle to reduce water loss.

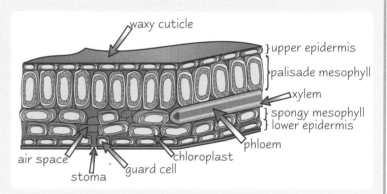

waxy cuticle
upper epidermis
palisade mesophyll
xylem
spongy mesophyll
lower epidermis
phloem
chloroplast
air space
stoma
guard cell

Plan Diagrams show Tissue Types in Organs

This is a **plan diagram** of a cross-section of a piece of leaf. Only **tissues** are shown on plan diagrams, **not cells**.

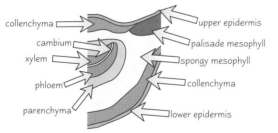

collenchyma
cambium
xylem
phloem
parenchyma
upper epidermis
palisade mesophyll
spongy mesophyll
collenchyma
lower epidermis

Remember, the **linear magnification** of drawings like this is **calculated as**:

$$\frac{\text{length of drawing or photograph}}{\text{length of specimen object}}$$

Practice Questions

Q1   What is the definition of a tissue?

Q2   Name one place in the human body where you would find ciliated epithelium.

Q3   What three tissues is the liver organ composed of?

Exam Questions

Q1   Give two examples of epithelial tissue found in mammals and, for each, explain how their structure suits them to their function.   [6 marks]

Q2   Explain how each of the tissues in a leaf is adapted to allow the leaf to perform its function.   [12 marks]

Cells — tiny little blobs with important jobs to do...

Make sure you learn the examples of tissues and organs and how cells are adapted to their jobs — examiners love examples. And make sure you know the differences between eukaryotic and prokaryotic cells. You may think I'm a bossy old witch, and you'd be right. But if you want to pass your exams, you'll have to do what I say. Oooh, the power, the power...

Carbohydrates

All carbohydrates contain only carbon, hydrogen and oxygen. Carbohydrates are dead important chemicals — for a start they're the main energy supply in living organisms and some of them, like cellulose, have an important structural role.

Carbohydrates are Made from **Monosaccharides**

All carbohydrates are made from sugar molecules. A single sugar molecule is called a **monosaccharide**. Examples of monosaccharides include glucose, fructose, ribose, deoxyribose and galactose.

There are two types of glucose — **alpha** (α) and **beta** (β) glucose. You need to know how their molecules are arranged slightly differently. This has important effects on their **properties** and **functions** (see p.9).

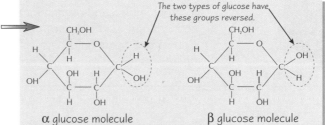

The two types of glucose have these groups reversed.

α glucose molecule β glucose molecule

Remember, beta glucose has the H on the bottom as you look at the structural diagram.

Condensation Reactions *Join Sugars Together*

Monosaccharides join to form larger carbohydrates by forming **glycosidic bonds**. When this happens, a molecule of water is squeezed out. This is called a **condensation reaction**.

If you're asked to show a condensation reaction in an exam, don't forget to put the water molecule in as a product.

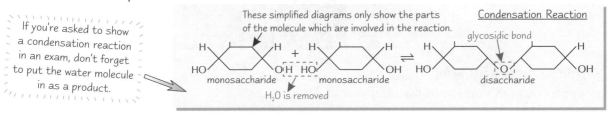

These simplified diagrams only show the parts of the molecule which are involved in the reaction.

Condensation Reaction

monosaccharide   monosaccharide   glycosidic bond   disaccharide

H_2O is removed

Hydrolysis Reactions *Break Sugars Apart*

When sugars are separated, the condensation reaction goes into **reverse**. This is called a **hydrolysis reaction** — a water molecule reacts with the glycosidic bond and **breaks it apart**.

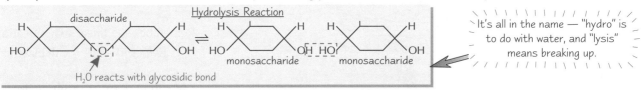

disaccharide   Hydrolysis Reaction   monosaccharide   monosaccharide

H_2O reacts with glycosidic bond

It's all in the name — "hydro" is to do with water, and "lysis" means breaking up.

Condensation and hydrolysis reactions are dead important in biology. **Proteins and lipids** are put together and broken up by them as well. So you definitely need to understand how they work.

The **Benedict's Test** tests for **Sugars**

The Benedict's test identifies **reducing sugars**. These are sugars that can donate electrons to other molecules — they include **all monosaccharides** and **some disaccharides** (two monosaccharides joined together, e.g. maltose).
When added to reducing sugars and heated, the **blue Benedict's reagent** gradually turns **brick red** due to the formation of a **red precipitate**.

The colour changes from:

blue — green — yellow — orange — brick red.

The higher the concentration of reducing sugar, the further the colour change goes — you can use this to **compare** the amount of reducing sugar in different solutions. A more accurate way of doing this is to **filter** the solution and **weigh the precipitate**.

boiling tube

beaker

test solution and Benedict's reagent

water

gauze

tripod

heat

To test for **non-reducing sugars** like sucrose (which is a **disaccharide**), you first have to break them down chemically into monosaccharides. You do this by boiling the test solution with **dilute hydrochloric acid** and then neutralising it with sodium hydrogen carbonate before doing the Benedict's test.

Carbohydrates

Polysaccharides are Loads of Sugars Joined Together

Polysaccharides are molecules which are made up of **loads of sugar molecules** stuck together.
You need to know about :

1) **starch** — the main storage material in plants;

2) **glycogen** — the main storage material in animals;

3) **cellulose** — the major component of cell walls in plants.

Examiners like to ask about the link between the structures of polysaccharides and their functions.

Amylose

①**Starch** is made up of **two** other polysaccharides of **alpha-glucose**:
- **Amylose** is a long, **unbranched chain** of alpha-glucose. The angles of the glycosidic bonds give it a **coiled structure**, almost like a cylinder. Its **compact**, coiled structure makes it really **good for storage**.
- **Amylopectin** is a long, **branched chain** of alpha-glucose. Its **side branches** make it particularly good for the storage of glucose — the enzymes that break down the molecule can get at the glycosidic bonds easily, to break them and release the glucose.

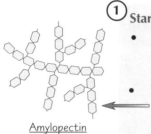
Amylopectin

②**Glycogen** is a polysaccharide of **alpha-glucose**. Its structure is very similar to amylopectin, except that it has **loads** more **side branches** coming off it. It's a very **compact** molecule found in animal liver and muscle cells. Loads of branches mean that stored glucose can be released quickly, which is **important for energy release** in animals.

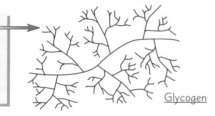
Glycogen

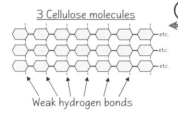
3 Cellulose molecules

Weak hydrogen bonds

③**Cellulose** is made of long, unbranched chains of **beta-glucose**. The bonds between beta sugars are **straight**, so the chains are straight. The chains are linked together by **hydrogen bonds** (see p.13) to form strong fibres called **microfibrils**. The strong fibres mean cellulose can provide **structural support** for cells. Another feature is that the **enzymes** that break the glycosidic bonds in starch can't reach the glycosidic bonds in cellulose, so those enzymes **can't break down cellulose**.

The Iodine Test tests for Starch

In this test, you don't have to make a **solution** from the substance you want to test — you can use **solids** too. Dead easy — just add **iodine dissolved in potassium iodide solution** to the test sample. If there's starch present, the sample changes from **browny-orange** to a dark, **blue-black** colour.

Practice Questions

Q1   What is the name given to the type of bond that holds sugar molecules together?

Q2   How can you work out the different concentrations of reducing sugars in two solutions?

Q3   Name the two different types of molecule that are combined together in a starch molecule.

Q4   Cellulose is made from beta glucose. How does this help with its function as a structural polysaccharide?

Exam Questions

Q1   Describe how glycosidic bonds in carbohydrates are formed and broken in living organisms.   [7 marks]

Q2   Compare and contrast the structures of glycogen and cellulose,
showing how each molecule's structure is linked to its function.   [10 marks]

Who's a pretty polysaccharide, then...

If you learn these basics it makes it easier to learn some of the more complicated stuff later on — 'cos carbohydrates crop up all over the place in biology. Remember that condensation and hydrolysis reactions are the reverse of each other — and don't forget that starch is composed of two polysaccharides. So many reminders, so little space ...

Lipids

*Lipids are fats, oils and waxes — they are all made up of carbon, hydrogen and oxygen,
and they're all insoluble in water. Ever seen a candle dissolve in water? No — exactly.*

Most Fats and Oils are **Triglycerides**

Most lipids are composed of compounds called triglycerides. Triglycerides are
composed of one molecule of **glycerol** with **three fatty acids** attached to it.

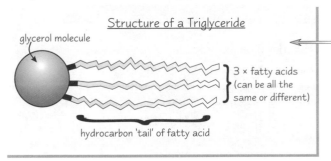

Structure of a Triglyceride

glycerol molecule

3 × fatty acids
(can be all the
same or different)

hydrocarbon 'tail' of fatty acid

Fatty acid molecules have long 'tails' made of **hydrocarbons**.
The tails are '**hydrophobic**' (they repel water molecules).
These tails make lipids insoluble in water. When put in water,
fat and oil molecules **clump together** in globules to reduce
the surface area in contact with water.

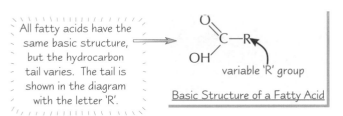

All fatty acids have the
same basic structure,
but the hydrocarbon
tail varies. The tail is
shown in the diagram
with the letter 'R'.

variable 'R' group

Basic Structure of a Fatty Acid

Triglycerides are **Formed** by **Condensation Reactions**

Like carbohydrates, lipids are formed by **condensation reactions** and broken up by **hydrolysis reactions**.
The diagram below shows a **fatty acid** joining to a **glycerol molecule**, forming an **ester bond**.
A molecule of water is also formed — it's a **condensation reaction**. This process happens twice more, to
form a **triglyceride**. The **reverse** happens in **hydrolysis** — a molecule of water is added to each ester bond to
break it apart, and the triglyceride splits up into three fatty acids and one glycerol molecule.

Formation of a Triglyceride

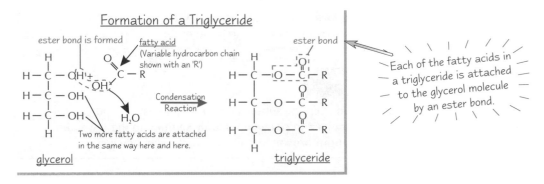

Each of the fatty acids in
a triglyceride is attached
to the glycerol molecule
by an ester bond.

Phospholipids are a Special Type of Lipid

The lipids found in **cell membranes** aren't triglycerides — they're **phospholipids**.
The difference is small but important:

1) In phospholipids, a **phosphate group**
   replaces one of the fatty acid molecules.

2) The phosphate group is **ionised**, which
   makes it **attract water** molecules.

3) So part of the phospholipid molecule is
   hydrophilic (attracts water) while the rest
   (the fatty acid tails) is **hydrophobic** (repels
   water). This is important in the cell
   membrane (see p.20 to find out why).

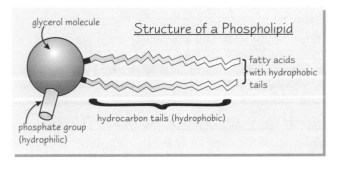

glycerol molecule

Structure of a Phospholipid

fatty acids
with hydrophobic
tails

hydrocarbon tails (hydrophobic)

phosphate group
(hydrophilic)

Lipids

Lipids are Useful

1) Lipids contain a lot of **energy per gram**, so they make useful **medium** or **long-term energy stores**. But they can't be broken down very quickly, so organisms use carbohydrates for **short-term storage**.

2) Lipids stored under the skin in **mammals** act as **insulation**. Skin loses heat from blood vessels, but the fatty tissue under the skin doesn't have an extensive blood supply, so it conserves heat.

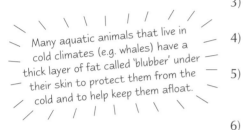

Many aquatic animals that live in cold climates (e.g. whales) have a thick layer of fat called 'blubber' under their skin to protect them from the cold and to help keep them afloat.

3) In **marine mammals** (e.g. whales, seals) lipids provide **buoyancy**, because the density of lipids is lower than that of muscle and bone.

4) Lipids under the skin and around the internal organs also provide **physical protection**, acting as a **cushion** against blows.

5) Lipids can act as a **waterproofing** layer — for example in the **waxy cuticle** on the surface of leaves and in the **exoskeleton** of insects. Lipids don't mix well with water, so water can't get through a lipid layer very easily.

6) Lipids provide animals with a **source of water** when they're respired. This is especially important for animals that live in the desert.

The Emulsion Test Tests for Lipids

Shake the test substance with **ethanol** for about a minute, then pour the solution into water. Any lipid will show up as a **milky emulsion**. The more lipid there is, the more noticeable the milky colour will be.

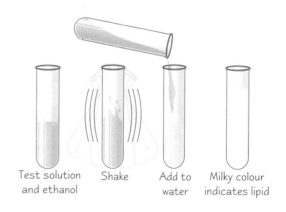

Test solution and ethanol | Shake | Add to water | Milky colour indicates lipid

Practice Questions

Q1 Why are lipids insoluble in water?

Q2 Explain the difference between a triglyceride and a phospholipid.

Q3 How can you test whether a lipid is present?

Q4 Give three ways that lipids are useful to living organisms.

Exam Questions

Q1 State five functions of lipids in animals. For each, explain the feature of lipids that allows them to perform this function. [10 marks]

Q2 Describe the differences between a triglyceride and a phospholipid, and explain how these differences affect the properties of the molecule. [8 marks]

What did the seal say to the upset whale? — Quit blubbering...

Truly awful joke — I hang my head in shame. You don't get far in life without extensive lard knowledge, so learn all the details on this page good and proper. Lipids pop up in other sections, so make sure you know the basics about how their structure gives them some quite groovy properties. Right, all this lipids talk is making me hungry — chips time...

Proteins

There are hundreds of different proteins — all of them contain carbon, hydrogen, oxygen and nitrogen. They are the most abundant organic molecules in cells, making up 50% or more of a cell's dry mass — now that's just plain greedy.

Proteins are Made from Long Chains of Amino Acids

All proteins are made up of amino acids joined together. All amino acids have a **carboxyl group (-COOH)** and an **amino group (-NH$_2$)** attached to a carbon atom.

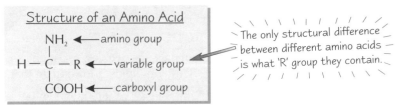

Structure of an Amino Acid

NH$_2$ ◄—— amino group

H — C — R ◄—— variable group

COOH ◄—— carboxyl group

The only structural difference between different amino acids is what 'R' group they contain.

Proteins are Formed by Condensation Reactions

Just like carbohydrates and lipids, the parts of a protein are put together by **condensation** reactions and broken apart by **hydrolysis** reactions. The bonds that are formed between amino acids are called **peptide bonds**.

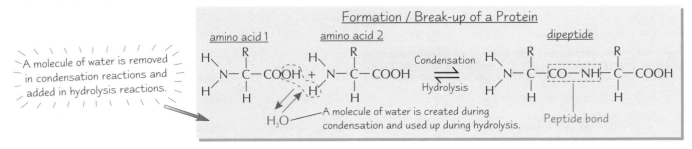

Formation / Break-up of a Protein

A molecule of water is removed in condensation reactions and added in hydrolysis reactions.

A molecule of water is created during condensation and used up during hydrolysis.

Peptide bond

Proteins have up to Four Structures

Proteins are **big, complicated** molecules. They're easier to explain if you describe their structure in four 'levels'. These levels are called the protein's **primary, secondary, tertiary** and **quaternary** structures.

① The **primary structure** is the **sequence of the amino acids** in the long chain that makes up the protein (the **polypeptide chain**).

COOH—Leucine—Arginine—Cysteine—Glycine—Arginine

amino acids

more amino acids (not drawn)

free COOH group

Glycine—Phenylalanine—Lysine—Valine—NH$_2$

free NH$_2$ group

② The **bonds** between the amino acids make the chain form a sort of **coil**. The way the chain coils is called its **secondary structure**. The most common secondary structure is a **spiral** called an **alpha (α) helix**.

α helix chain

③ The coiled chain of amino acids is itself often coiled and folded in a characteristic way that identifies the protein. **Extra bonds** can form between different parts of the polypeptide chain, which gives the protein a kind of **three dimensional shape**. This is its **tertiary structure**.

α helix chain coiled into tertiary structure

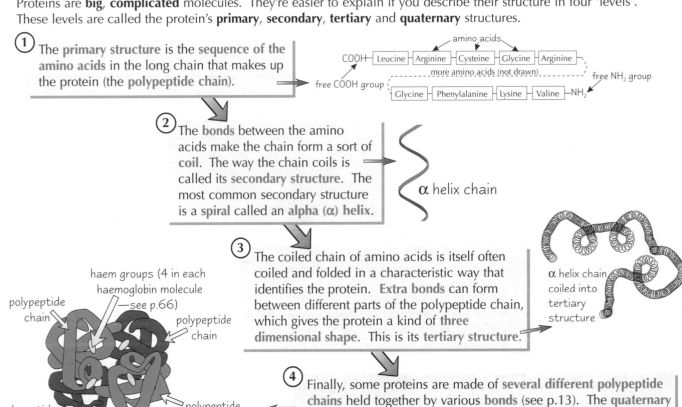

haem groups (4 in each haemoglobin molecule —see p.66)

polypeptide chain

polypeptide chain

polypeptide chain

polypeptide chain

A haemoglobin molecule

④ Finally, some proteins are made of **several different polypeptide chains** held together by various **bonds** (see p.13). The **quaternary structure** is the way these different parts are assembled together.

Proteins

Different Bonds Hold Proteins Together

Various types of bond hold protein molecules in shape.

Hydrogen bonds are weak bonds formed when a hydrogen atom is attracted by a slight negative charge on another atom.

1) **Hydrogen bonds** hold together the **secondary** structure of a protein. For example, in an alpha helix, a hydrogen bond forms between the **C=O** group of one amino acid and the **N-H** group of **another amino acid**, four amino acids along the polypeptide chain. Hydrogen bonds also help hold the **tertiary** structure of a protein together.

2) The **tertiary** structure of a protein is also held together by **weak ionic bonds**. These are weak attractions between a negatively charged part of one molecule and a positively charged part of another.

3) Whenever two molecules of the amino acid **cysteine** come close together, the sulphur in one cysteine bonds to the sulphur in the other cysteine. This is called a **disulphide bond**. It's part of the tertiary structure of a protein.

4) When water-repelling **hydrophobic** groups are close together in the protein, they tend to **clump closely together**. These **hydrophobic bonds** are important in making the protein **fold up** into its final structure.

Protein Shape Relates to its Function

You need to learn **two examples** of how proteins are **adapted for their jobs**.

Haemoglobin is a **globular protein** that absorbs oxygen (see diagram on p.12). It has a complex tertiary structure which is curled up, so **hydrophilic** ('water-attracting') side chains are on the **outside** of the molecule and **hydrophobic** ('water-repelling') side chains face **inwards**. This makes haemoglobin **soluble in water** and **easily transported** in the blood. The four polypeptide chains in haemoglobin support four **haem** groups. These are not proteins — each haem group is made up of four **iron ions** (Fe^{2+}) which carry the oxygen molecules.

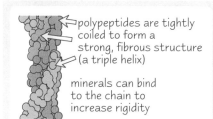

polypeptides are tightly coiled to form a strong, fibrous structure (a triple helix)

minerals can bind to the chain to increase rigidity

Collagen is a **fibrous** protein that forms **supportive tissue** in animals, so it has to be strong.

It's insoluble, and made up of three long polypeptide chains coiled tightly together.

The Biuret Test tests for Proteins

There are **two stages** to this test.

1) The test solution needs to be **alkaline**, so first you add a few drops of **2M sodium hydroxide**.

2) Then you add some **0.5% copper (II) sulphate solution**. If a **purple layer** forms, there's protein in it. If it stays **blue**, there isn't. The colours are pale, so you need to look carefully.

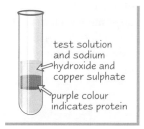

test solution and sodium hydroxide and copper sulphate

purple colour indicates protein

Practice Questions

Q1 What are the common features of all amino acid molecules?

Q2 What is the name given to the bond that holds amino acids together in proteins?

Q3 What are the four types of bond that hold a protein molecule in shape?

Exam Questions

Q1 Describe the structure of a protein, explaining the terms primary, secondary, tertiary and quaternary structure. No details are required of the chemical nature of the bonds. [9 marks]

Q2 Describe the structure of the collagen molecule, and explain how this structure relates to its function in the body. [6 marks]

The name's Bond — Peptide Bond...

Quite a lot to learn on these pages — proteins are annoyingly complicated. Not happy with one, or even two, structures — they've got four of the things, and you need to learn 'em all. Condensation and hydrolysis reactions are back by popular demand and you need to learn all the different bonds too.

Water and Inorganic Ions

*Life can't exist without water — in fact boring, everyday water is one
of the most important substances on the planet. Funny old world.*

Water is Vital to Living Organisms

Water makes up about 80% of cell contents — it has loads of important **functions**, inside and outside cells.

1) Water is a **metabolic reactant**. That means it's needed for loads of important
chemical reactions, like photosynthesis and hydrolysis reactions (remember them...).

2) Water is a **solvent** — which means it can dissolve many substances.
Most biological reactions take place **in solution**, so water's solvent properties are vital.

3) Water **transports** substances. The fact that it's a **liquid** and a solvent makes it easy for water
to transport all sorts of materials around plants and animals, like glucose and oxygen.

4) Water helps with **temperature control**. When water **evaporates**, it uses up heat
from the surface that it's on. This cools the surface and helps lower the temperature.

Water Molecules have a Simple Structure

The **structure of a water molecule** helps to explain many of its **properties**.
Examiners like asking you to relate structure to properties, so make sure you're clear on this.

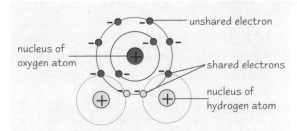

Water is **one atom of oxygen** joined to **two atoms of
hydrogen** by **shared electrons**. Because the shared
hydrogen electrons are pulled close to the oxygen atom,
the other side of each hydrogen atom is left with a **slight
positive charge**. The unshared electrons on the oxygen
atom give it a **slight negative charge**. That means water is a
polar molecule — it has negative charge on one side and
positive charge on the other.

The **negatively charged oxygen atoms** of water
attract the **positively charged hydrogen atoms** of
other water molecules. This attraction is called
hydrogen bonding and it gives water some of its
special properties.

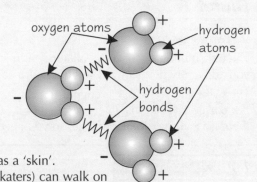

1) **Cohesion** — water molecules tend to stick together.
This property enables water to flow.

2) **Surface tension** — water behaves as though it has a 'skin'.
That's why some small invertebrates (like pond-skaters) can walk on
water. It wouldn't happen if the molecules weren't held together.

3) **High specific heat capacity** — specific heat capacity is the **energy required**
to raise the temperature of 1 gram of a compound by 1°C. Water has a
high specific heat capacity — it takes a lot of energy to heat it up. This is
useful for aquatic organisms, as it stops rapid temperature changes.

4) **High latent heat of evaporation** — it takes a lot of
heat to evaporate water, so it's great for cooling things.

Ice Floats — Which is Useful

Another weird property of water that's useful in nature is that it reaches its **maximum density** at about **4°C**.
This means that ice is **less dense** than the water around it, so it **floats**. This acts as an **insulation** for the water
below — so the sea or a lake won't freeze solid, which allows organisms under the ice to survive.

Water and Inorganic Ions

Water's *Polarity* Makes it a *Solvent*

Because water is **polar**, it's a **good solvent** for other polar molecules. **Ionic** substances like salt, and **organic** molecules that have an **ionised group** will dissolve in water. The **positive** end of a water molecule will be attracted to a **negative ion** and the negative end of a water molecule will be attracted to a **positive ion**. The ion gets **totally surrounded** by water molecules — in other words, it **dissolves**.

Remember — a molecule is polar if it has a negatively charged bit and a positively charged bit. This is also called "uneven charge distribution".

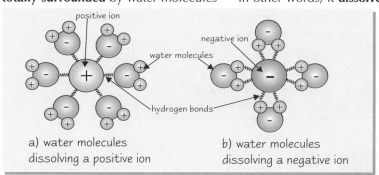

a) water molecules dissolving a positive ion

b) water molecules dissolving a negative ion

Chemical reactions take place much more easily in solution, because the dissolved ions are more **free to move around** and react than if they were held tightly together in a solid. Once dissolved, the solute can **easily be transported** by the water.

Inorganic Ions are Important for *Metabolism*

Ions are **charged particles** — **inorganic** ions are those that **don't contain carbon**. Many **inorganic ions** that can dissolve in water are important in metabolism. Some of the most important inorganic ions and their uses are shown below:

ION	IMPORTANT USE
Calcium (Ca^{2+})	Important in the formation of bone.
Sodium (Na^+)	Involved in nerve transmission.
Potassium (K^+)	Involved in activating enzymes.
Magnesium (Mg^{2+})	Contained in chlorophyll.
Chloride (Cl^-)	Used to produce hydrochloric acid in the stomach.
Nitrate (NO_3^-)	Needed for the synthesis of proteins by plants.
Phosphate (PO_4^{3-})	Needed for the formation of ATP.

Practice Questions

Q1   State four uses of water in living organisms.

Q2   What is a 'polar molecule'?

Q3   Explain why it is useful for aquatic organisms that water has a high specific heat capacity.

Q4   What is phosphate needed for?

Exam Questions

Q1   Relate the structure of the water molecule to its uses in living organisms. [12 marks]

Q2   Plants deprived of the mineral magnesium tend to be unhealthy and their leaves are yellow, whilst those deprived of nitrogen have stunted growth. Suggest reasons for these observations. [7 marks]

Pssssssssssssssssssssssssssssssssssssssssssssss — need the loo yet?

Water is pretty darn useful really. It looks so, well, dull— but in fact it's scientifically amazing, and essential for all kinds of jobs — like maintaining aquatic temperatures, transporting things and enabling reactions. You need to learn all its properties and uses, plus the uses of inorganic ions. Right, I'm off — when you gotta go, you gotta go.

Action of Enzymes

*Enzymes crop up loads in biology — they're really useful 'cos they make reactions work more quickly. So, whether you feel the need for some speed or not, read on — because you **really** need to know this basic stuff about enzymes.*

Enzymes are Biological Catalysts

Enzymes speed up chemical reactions by acting as **biological catalysts**.

1) They catalyse every **metabolic reaction** in the bodies of living organisms. Even your **phenotype** (physical appearance) is down to enzymes that catalyse the reactions that cause growth and development.

2) Enzymes are **globular proteins** (see p.13) although some have **non-protein components** too.

3) Every enzyme has an area called its **active site**. This is the part that connects the enzyme to the substance it interacts with, which is called the **substrate**.

Enzymes Reduce Activation Energy

In a chemical reaction, a certain amount of energy needs to be supplied to the chemicals before the reaction will start. This is called the **activation energy** — it's often provided as **heat**. Enzymes **reduce** the amount of activation energy that's needed, often making reactions happen at a **lower temperature** than they could without an enzyme. This **speeds** up the **rate of reaction**.

When a substrate fits into the enzyme's active site it forms an **enzyme-substrate complex**:

1) If two substrate molecules need to be **joined**, attaching to the enzyme holds them **close together**, **reducing** any **repulsion** between the molecules so they can bond more easily.

2) If the enzyme is catalysing a **breakdown reaction**, fitting into the active site puts a **strain** on bonds in the substrate, so the substrate molecule **breaks up** more easily.

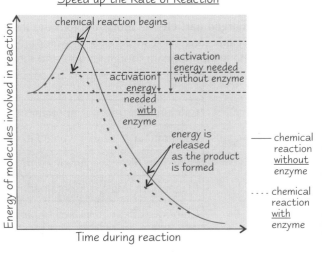

Graph Showing How Enzymes Speed up the Rate of Reaction

Enzymes are a bit picky. They only work with **specific substrates** — usually only one. This is because, for the enzyme to work, the substrate has to **fit** into the **active site**. This property of enzymes is known as **enzyme specificity**.

Temperature has a Big Influence on Enzyme Activity

Like any chemical reaction, the rate of an enzyme-controlled reaction increases when the temperature's raised. More heat means more **kinetic energy**, so molecules move faster. This makes the enzyme more likely to **collide** with the substrate. But, if the temperature increases beyond a certain point, the **reaction stops**. This is because the rise in temperature also makes the enzyme's particles **vibrate**:

1) If the temperature goes above a certain level, this vibration **breaks** some of the **bonds** that hold the enzyme in shape.

2) The **active site changes shape** and the enzyme and substrate **no longer fit together**.

3) At this point, the enzyme is **denatured** — it no longer functions as a catalyst.

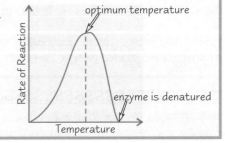

Action of Enzymes

pH *Also Affects Enzyme Activity*

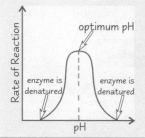

All enzymes have an **optimum pH value**. Most work best at neutral pH 7, but there are exceptions. **Pepsin**, for example, works best at acidic pH 2, which suits it to its role as a stomach enzyme. Above and below the optimum pH, the H+ and OH- ions found in acids and alkalis can mess up the **ionic bonds** that hold the enzyme's tertiary structure in place. This makes the active site change shape, so the enzyme is **denatured**.

Enzyme Concentration Affects the Rate of Reaction

1) The **more enzyme molecules** there are in a solution, the more likely a substrate molecule is to **collide** with one. So increasing the concentration of the enzyme increases the rate of reaction.

2) But if the amount of substrate is limited, there comes a point when there's more than enough enzyme to deal with all the available substrate, so adding more enzyme has **no further effect**.

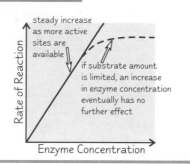

Substrate Concentration Affects the Rate of Reaction **Up To a Point**

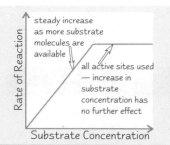

Substrate concentration affects the rate of reaction up to a certain point. The higher the substrate concentration, the faster the reaction, but only up until a **'saturation' point**. After that, there's so many substrate molecules that the enzymes have about as much as they can cope with, and adding more **makes no difference**.

Practice Questions

Q1 Define the term "catalyst".

Q2 What is the name given to the amount of energy needed to start a reaction?

Q3 What is an "enzyme-substrate complex"?

Q4 Explain why enzymes are specific (i.e. only work with a single or a small group of substrates).

Q5 Why do high temperatures denature enzymes?

Q6 Explain why increasing the concentration of an enzyme doesn't always increase the rate of reaction.

Exam Question

Q1 When doing an experiment on enzymes, explain why it is necessary to control the temperature and pH of the solutions involved.

[8 marks]

But why is the enzyme-substrate complex?

Don't let all those graphs put you off — there's not really anything very hard on this page. The main thing to understand is that enzymes need quite a specific set of conditions to work at their best, and even small differences in temperature or pH can lead to a big drop in their performance. Skivers.

Enzyme Activity

Just when you thought you'd seen the last of enzymes, here they are again to brighten up your day one more time. Don't worry, there's not that much to learn on these two pages.

There are two main ways of measuring the rate of an enzyme-controlled reaction:

1 You can **Measure** How Fast the **Product** of the Reaction **Appears**

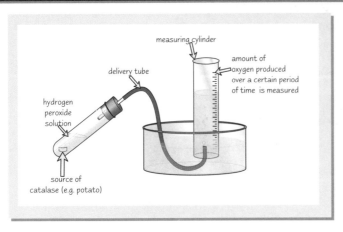

The diagram shows how to measure the rate of a reaction controlled by the enzyme **catalase**. Catalase is found in many living organisms — it catalyses the **breakdown** of **hydrogen peroxide** into **water** and **oxygen**. It's easy to collect the oxygen produced and measure **how fast** it's given off.

Below is a typical graph of results for an experiment like this one:

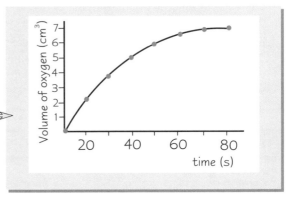

The graph shows that the most oxygen is produced at the beginning of the reaction. The rate of the reaction is quicker at first because there's a high concentration of substrate. As it's used up, the rate of the reaction gets slower and finally the reaction stops. This is where the graph levels off — no more oxygen is produced.

2 But Sometimes it's Easier to Measure **Disappearance** of **Substrate** Instead

For example, **amylase** enzyme catalyses the **breakdown** of **starch** to **maltose**. Starch is easier to detect than maltose, so it's easier to time the **disappearance** of the starch. The diagram below shows the method:

You could use a **colorimeter** to give a more accurate idea of how much starch is present. This measures the **absorbance** of samples taken from the reaction mixture. The darkest blue-black colour (showing lots of starch is present) gives the highest value for absorbance. Which is handy.

Here's a typical graph of results for this type of experiment:

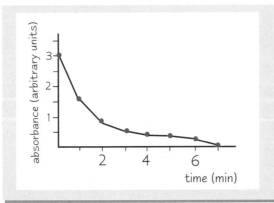

The graph shows that the absorbance readings decrease quickly at first and then more slowly. This is because there are a lot of starch molecules at first, but then there are fewer and fewer left to be broken down as the reaction continues.

Enzyme Activity

Enzyme Activity can be Inhibited

Enzyme activity can be prevented by **enzyme inhibitors** — molecules that **bind to the enzyme** that they inhibit. Inhibition can be **competitive** or **non-competitive**.

Competitive inhibitors have a **similar shape to the substrate**. They compete with the substrate to bind to the active site, but no reaction follows. Instead they **block** the active site, so **no substrate** can **fit** in it. How much inhibition happens depends on the **relative concentrations** of inhibitor and substrate — if there's a lot of the inhibitor, it'll take up all the active sites and stop any substrate from getting to the enzyme.

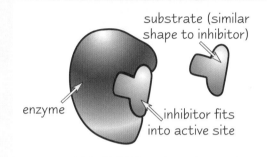

substrate (similar shape to inhibitor)

enzyme

inhibitor fits into active site

Non-competitive inhibitors bind to the enzyme **away from its active site**, but this causes the active site to **change shape**. They don't 'compete' with the substrate because even if there's a substrate in the active site, the inhibitor can still fit on.

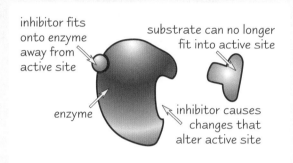

inhibitor fits onto enzyme away from active site

substrate can no longer fit into active site

enzyme

inhibitor causes changes that alter active site

Competitive and non-competitive inhibitors can be **reversible** or **non-reversible**. This mainly depends on the **strength of the bond** between the enzyme and the inhibitor.

1) If it's a **strong, covalent bond** then the inhibitor can't be removed easily and the inhibition is **irreversible**.

2) If it's a **weaker hydrogen bond** or a weak, **intermolecular ionic bond**, then the inhibitor can be removed and the inhibition is **reversible**.

Practice Questions

Q1   State the two things you can measure to establish the rate of an enzyme-controlled reaction.

Q2   For each of these methods, give an example of a particular enzyme reaction you could follow.

Q3   What is the difference between a competitive and a non-competitive enzyme inhibitor?

Q4   What is the difference between a reversible and an irreversible enzyme inhibitor?

Exam Questions

Q1   Catalase is an enzyme that catalyses the breakdown of hydrogen peroxide into oxygen and water.
Explain how you could measure the rate of this reaction, using potato as the source of enzyme.   [4 marks]

Q2   When a small amount of chemical X is added to a mixture of an enzyme and its substrate,
the formation of reaction products is reduced. Increasing the amount of X in the solution
causes further reduction in products. State, with reasons, the likely nature of chemical X.   [4 marks]

Don't be shy — lose your inhibitions and learn these pages...

It's not easy being an enzyme. They're just trying to get on with their jobs, but the whole world seems to be against them sometimes. High temperature, wrong pH, and now inhibitors — all out to get them. Sad though it is, make sure you know every word. Go on, it won't take long and it's really important stuff.

The Cell Membrane Structure

Two pages all about cell membranes and what they're made of. Try and contain your excitement when you read about the fluid mosaic structure — there have been some nasty cases of extreme over-excitement in the past.

Cell Membranes have a 'Fluid Mosaic' Structure

The **structure** of all **membranes** is basically the same. They are composed of **lipids** (mainly phospholipids), **proteins** and **carbohydrates** (usually attached to proteins or lipids).

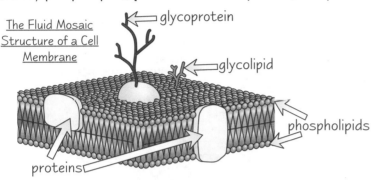

The Fluid Mosaic Structure of a Cell Membrane

glycoprotein

glycolipid

phospholipids

proteins

In 1972, the **fluid mosaic model** was suggested to describe the arrangement of molecules in the membrane. In the model, **phospholipid molecules** form a continuous, double layer (**bilayer**). This layer is 'fluid' because the phospholipids are constantly moving. **Protein molecules** are scattered through the layer, like tiles in a **mosaic**.

Detailed 3D pictures of cell membranes support the fluid mosaic model. Also, experiments with cell fusion show that proteins move about in the membrane, which means that the membrane is fluid.

Phospholipids Can Form Bilayers

Phospholipids consist of a **glycerol molecule** plus **two molecules** of **fatty acid** and a **phosphate group** (see p.10).

1) The phosphate / glycerol head is **hydrophilic** — it attracts water.
   The **fatty acid tails** are **hydrophobic** — they repel water.

2) In **aqueous (water-based) solutions** phospholipids automatically arrange themselves into a **double layer** so that the **hydrophobic tails** pack together **inside the layer** away from the water, and the **hydrophilic heads face outwards** into the aqueous solutions. This is the basis of the membrane.

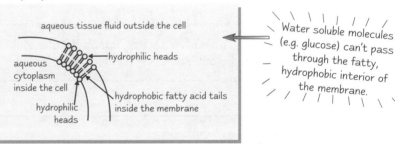

aqueous tissue fluid outside the cell

hydrophilic heads

aqueous cytoplasm inside the cell

hydrophilic heads

hydrophobic fatty acid tails inside the membrane

Water soluble molecules (e.g. glucose) can't pass through the fatty, hydrophobic interior of the membrane.

Cholesterol Gives the Membrane More Stability

1) **Cholesterol** belongs to a group of lipids called **steroids**. It's present in all cell membranes except those of bacteria. It can make up to 25% of the lipids in **animal cell membranes** but isn't found so much in plant membranes.

2) Having cholesterol molecules between phospholipid molecules makes the membrane **less fluid** and **more stable**.

Glycolipids and Glycoproteins Contain Polysaccharides

1) **Glycolipids** are lipids that have **combined with polysaccharides**. They're found in the **outer layer** of cell membranes. Their exact role isn't known but they may be involved in **cell recognition**.

2) **Glycoproteins** (also found in the **outer layer**) are proteins with **attached polysaccharides** of short, branched **chains of monosaccharides**. Glycoproteins have a variety of specific shapes due to the **different branching patterns** of the monosaccharides. These allow different cells to recognise each other. For example, some glycoproteins are **antigens** — they're recognised by **white blood cells**, which starts an immune response (see p.58 for more on this).

Remember — polysaccharides are made up of chains of single sugar molecules, (monosaccharides).

The Cell Membrane Structure

Proteins Have **Different Functions** in the **Membrane**

1) **Channel proteins** form a tiny **gap** in the membrane to allow water soluble molecules and ions through by diffusion.

2) **Carrier proteins** carry water soluble molecules and ions through the membrane by **active transport** and **facilitated diffusion** (see p.24).

3) **Receptor proteins** recognise and bind to **specific molecules** (e.g. hormones).

4) **Enzymes** can be embedded in the inner membrane of a cell or organelle — e.g. **ATPase** in the inner membrane of mitochondria.

5) Proteins in the membrane also help **strengthen** the membrane. There are **hydrogen bonds** between the proteins and the hydrophilic heads of the phospholipids.

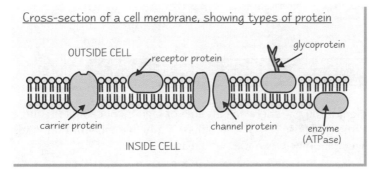

Cross-section of a cell membrane, showing types of protein

OUTSIDE CELL — receptor protein — glycoprotein

carrier protein — channel protein — enzyme (ATPase)

INSIDE CELL

Membranes Control What Passes Through Them

Cells and many of the **organelles** inside them are surrounded by **membranes**, which have a **range of functions**:

Membranes around organelles —

1) Divide the cell up into **different compartments** to make the different **functions more efficient** — e.g. the substances needed for **respiration** (like enzymes) are kept together inside **mitochondria**.

2) Provide a big surface area for **enzymes** or **pigments** that help the organelle do its job — e.g. **chlorophyll** pigment in **chloroplasts** (needed for photosynthesis).

3) Can be used to form **vesicles** to transport substances between different areas of the cell (e.g. see page 25).

Membranes around cells —

1) Allow **recognition** by other cells, e.g. the cells of the **immune system** (see page 58).

2) Provide **receptors** for molecules like **hormones**.

3) Act as a **barrier** to many substances.

4) Are **permeable** to small molecules like **oxygen** and **water**, and are **selectively permeable** to **ions** and larger molecules e.g. **glucose**.

Practice Questions

Q1 Give three functions of cell membranes.

Q2 Which types of molecule are carbohydrate molecules usually attached to in the cell membrane?

Q3 Name three molecules, other than phospholipids and proteins, that are present in animal cell membranes.

Exam Question

Q1 a) How does a phospholipid differ from a triglyceride? [1 mark]

b) Describe the role of phospholipids in controlling the passage of water soluble molecules through the cell membrane. [2 marks]

Membranes actually **are** all around...

The cell membrane is a complex structure — but then it has to be, 'cos it's the line of defence between a cell's contents and all the big bad molecules outside. Don't confuse the cell membrane with the cell wall (found in plant cells). The cell membrane controls what substances enter and leave the cell whereas the cell wall provides structural support.

Transport Across the Cell Membrane

There are six methods of transport across a cell membrane. You need to learn all six — diffusion, osmosis, facilitated diffusion, active transport, endocytosis and exocytosis. It's a big topic alright, sprawling over four whole pages. Oo-er.

1 Diffusion *is the* Passive Movement *of* Particles

1) If there's a **high concentration** of particles (molecules or ions) in one area of a liquid or gas, then these particles will gradually move and **spread out** into areas of **lower concentration**. Eventually, the particles will be **evenly distributed** throughout the liquid or gas. This movement is called diffusion.

2) Diffusion is described as a **passive process** because **no energy** is needed for it to happen.

3) Diffusion can happen **across cell membranes**, as long as the particles can **move freely** through the membrane. For example, water, oxygen and carbon dioxide molecules are small enough to pass easily through pores in the membrane.

The Speed of Diffusion *Depends on* Several Factors

1) The **concentration gradient** is the path between an area of higher concentration and an area of lower concentration. Particles diffuse **faster** when there is a **high concentration gradient** (a big difference in concentration between the two areas).

2) The **shorter** the **distance** the particles have to travel, the **faster** the rate of diffusion.

3) **Small molecules** move faster than large molecules, so they **diffuse faster**.

4) At **high temperatures** particles have more **kinetic** (movement) energy, so they **diffuse more quickly**.

5) The larger the **surface area** of the cell membrane, the faster the rate of diffusion.

Large *Organisms Maintain* Concentration Gradients *at* Exchange Surfaces

Substances diffuse from an area of **higher** concentration to an area of **lower** concentration. If there's a big difference in the concentration of a substance between two areas (a steep **concentration gradient**), the rate of diffusion is faster. One function of the **specialised exchange organs** of larger organisms is to maintain concentration gradients so things diffuse quickly. A good example of this is the **alveoli** in human lungs:

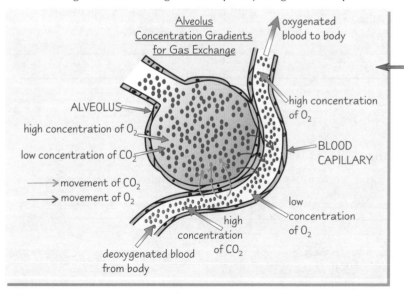

Alveolus Concentration Gradients for Gas Exchange

oxygenated blood to body

ALVEOLUS

high concentration of O_2

low concentration of CO_2

→ movement of CO_2

→ movement of O_2

high concentration of O_2

BLOOD CAPILLARY

high concentration of CO_2

low concentration of O_2

deoxygenated blood from body

After exchanging gases with cells, **deoxygenated blood** full of CO_2 returns to blood capillaries in the lungs. There's a higher level of O_2 in the **alveoli** compared to the capillaries. This gives a good concentration gradient for **diffusion of oxygen** into the **blood**. The low level of CO_2 in the **alveolar space** helps to remove the CO_2 from the blood. CO_2 **diffuses** down the concentration gradient and **into the lungs**, where it's breathed out.

Other exchange surfaces have similar features. E.g. blood delivers **glucose** to cells for respiration. When blood returns "empty" to the **microvilli** in the gut, more glucose diffuses down the **concentration gradient** from the gut into the **capillaries**. Most larger organisms have special ways of maintaining concentration gradients. See page 48 for more on the **features of exchange surfaces** that adapt them for exchange.

Transport Across the Cell Membrane

② *Osmosis is a Particular Kind of Diffusion*

1) Osmosis is when **water molecules** diffuse through a **partially permeable membrane** from an area of **higher water potential** (i.e. higher concentration of water molecules) to an area of **lower water potential**.

2) A **partially permeable membrane** allows some molecules through it, but not all.
Water molecules are small and can diffuse through easily but large solute molecules can't.

3) Water molecules will diffuse **both ways** through the membrane — but the **net movement** will be to the side with a **lower concentration of water molecules**.

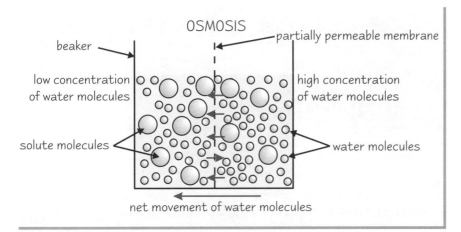

OSMOSIS

beaker → partially permeable membrane

low concentration of water molecules | high concentration of water molecules

solute molecules ← → water molecules

net movement of water molecules

Partially permeable membranes can be useful at sea.

D'oh Phew

Water Potential is the Ability of Water Molecules to Move

1) **Water potential** is the potential (likelihood) of water molecules to diffuse out of a solution.

> Water molecules are **more likely** to diffuse out of solutions with a **higher concentration** of water molecules. These solutions have a **high water potential**.
> Water molecules are **less likely** to diffuse out of solutions with a **lower concentration** of water molecules — these have a **low water potential**.

2) **Pure water** has the **highest water potential**. All solutions have a **lower** water potential than pure water.

Practice Questions

Q1 What is meant by the term "concentration gradient"?

Q2 What is a partially permeable membrane?

Q3 Define water potential.

Exam Question

Q1 The effect of temperature on the rate of diffusion of sodium ions was investigated. Pieces of potato, of equal surface area, were placed in distilled water at different temperatures. After 20 minutes the concentration of sodium ions in the water was measured and the results used to plot this graph.

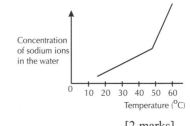

Concentration of sodium ions in the water

0   10 20 30 40 50 60
Temperature (°C)

Explain the increase in the rate of diffusion between 20°C and 50°C.

[2 marks]

*Warning — this topic may cause diffusion (I mean **con**fusion)...*

*A good way to describe **osmosis** is that it's the diffusion of water molecules through a partially permeable membrane from an area of **higher water potential** to an area of **lower water potential**. Water potential is a tricky idea — but it impresses the examiners, so try and get your head round it.*

Transport Across the Cell Membrane

③ Facilitated Diffusion uses Carrier Proteins and Channel Proteins

Some **larger molecules** (e.g. amino acids, glucose) and **charged atoms** (e.g. sodium ions) can't diffuse through the phospholipid bilayer of the cell membrane themselves. Instead they diffuse through **carrier proteins** or **channel proteins** in the cell membrane. This is called **facilitated diffusion**.

1) Channel proteins form **pores** through the membrane for charged particles to diffuse through.

2) Carrier proteins **change shape** to move large molecules into and out of the cell:

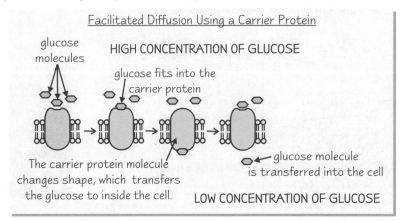

Facilitated Diffusion Using a Carrier Protein

glucose molecules

HIGH CONCENTRATION OF GLUCOSE

glucose fits into the carrier protein

The carrier protein molecule changes shape, which transfers the glucose to inside the cell.

glucose molecule is transferred into the cell

LOW CONCENTRATION OF GLUCOSE

The carrier proteins in the cell membrane have **specific shapes** — so specific carrier proteins can only facilitate the diffusion of specific molecules. Facilitated diffusion can only move particles along a **concentration gradient**, from a higher to a lower concentration. It **doesn't** use any **energy**.

④ Active Transport Moves Substances Against a Concentration Gradient

1) Active transport uses **energy** to move **molecules** and **ions** across cell membranes, **against** a **concentration gradient**.

2) Molecules attach to **specific carrier proteins** (sometimes called '**pumps**') in the **cell membrane**, then **molecules of ATP** (adenosine triphosphate) provide the energy to change the shape of the protein and move the molecules across the membrane.

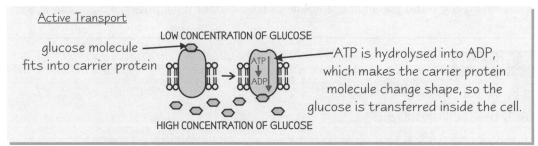

Active Transport

glucose molecule fits into carrier protein

LOW CONCENTRATION OF GLUCOSE

ATP is hydrolysed into ADP, which makes the carrier protein molecule change shape, so the glucose is transferred inside the cell.

HIGH CONCENTRATION OF GLUCOSE

Plants Take in Mineral Ions Using Active Transport

An example of active transport is the uptake of **mineral ions** into plant root hairs. Plants have to be able to take in minerals against a concentration gradient, because the concentration of a mineral the plant needs is often much greater than the concentration of that mineral in the soil. Plant **root hair cells** stick out into the soil, giving a **large surface area** for absorbing water and minerals from the soil. Their **cell membranes** include **carrier proteins** for the active transport of mineral ions into the plant. The cells also have lots of **mitochondria** to provide the **energy** for active transport.

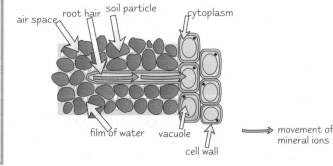

air space   root hair   soil particle   cytoplasm

film of water   vacuole   cell wall   movement of mineral ions

Transport Across the Cell Membrane

5) Materials can be **Taken into** Cells by **Endocytosis**

Endocytosis is when a cell takes in substances by surrounding them
with a section of the cell membrane to form a small vacuole called a **vesicle**.

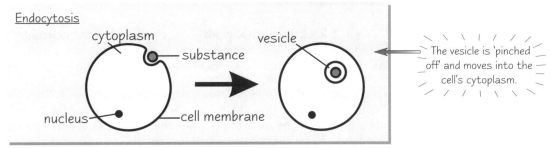

Endocytosis

cytoplasm — substance — vesicle

The vesicle is 'pinched off' and moves into the cell's cytoplasm.

nucleus — cell membrane

An example of endocytosis is **phagocytosis**, where sometimes whole cells are brought into the cell
(see page 58). The contents of the vesicle are then digested by enzymes from the lysosomes.

6) Materials can be **Removed from** Cells by **Exocytosis**

Materials are **secreted out** of cells by **exocytosis**.

1) Substances **produced by the cell** move through the
endoplasmic reticulum to the **Golgi apparatus**.

2) **Vesicles** pinch off from the sacs of the Golgi apparatus
and move towards the cell membrane.
They **merge** with the **cell membrane** and **release** their
contents outside of the cell.

3) Digestive enzymes, hormones, mucus and milk
are secreted by **exocytosis**.

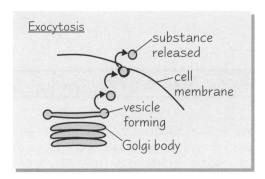

Exocytosis

substance released

cell membrane

vesicle forming

Golgi body

Practice Questions

Q1   Name the two types of protein used in facilitated diffusion.

Q2   Active transport and facilitated diffusion both involve carrier proteins.   Which one needs energy?

Q3   Why do plants need to take in minerals via active transport?

Q4   How does endocytosis differ from exocytosis?

Q5   Give two examples of substances secreted by exocytosis.

Exam Questions

Q1   Sometimes molecules are unable to diffuse through the phospholipid bilayer of the cell membrane.
Give three reasons why this might be, and briefly describe the ways in which such molecules
can be transported into and out of the cell.   [10 marks]

Q2   a)   In terms of water potential, explain how water moves from the soil into a root hair cell.   [3 marks]

b)   Explain how the special features of root hair cells increase the active uptake of ions.   [3 marks]

A little less conversation, a little more exocytosis, baby...

_Phew, the end of a mammoth topic on transport through the cell membrane — so now you can move on and forget it ever
happened. Just kidding (I should be doing stand-up, no really) — now you need to go back over it and check you know the
details. Learn the differences between similar terms, like exocytosis and endocytosis, and passive versus facilitated diffusion._

Structure of DNA and RNA

*These pages are about the structure of DNA (**deoxy**ribonucleic acid) and RNA (plain ol' ribonucleic acid), plus a little thing called the genetic code, which is kinda important to us living things. (OK, spot the major understatement here.)*

DNA and RNA are Very Similar Molecules

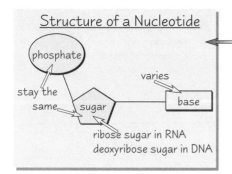

Structure of a Nucleotide

phosphate

stay the same

sugar

varies

base

ribose sugar in RNA
deoxyribose sugar in DNA

Although DNA is called deoxyribonucleic acid, it still contains oxygen.

DNA and RNA are **nucleic acids** — made up of lots of **nucleotides** joined together. Nucleotides are units made from a **pentose sugar** (with 5 carbon atoms), a **phosphate** group and a **base** (containing nitrogen and carbon).

The sugar in **DNA** nucleotides is a **deoxyribose** sugar — in **RNA** nucleotides it's a **ribose** sugar. Within DNA and RNA, the sugar and the phosphate are the same for all the nucleotides. The only bit that's different between them is the **base**. There are five possible bases and they're split into two groups:

	GROUP	BASIC STRUCTURE	BASE	found in DNA	found in RNA
1)	Purine bases	2 rings of atoms	adenine	✓	✓
			guanine	✓	✓
2)	Pyrimidine bases	single ring of atoms	cytosine	✓	✓
			thymine	✓	✗
			uracil	✗	✓

DNA and RNA are Polymers of Mononucleotides

Mononucleotides (single nucleotides) join together by a **condensation reaction** between the **phosphate** of one group and the **sugar** molecule of another. As in all condensation reactions, **water** is a by-product.

DNA is made of **two strands of nucleotides**. RNA has just the one strand. In DNA, the strands spiral together to form a **double helix**. The two strands are held together by **hydrogen bonds** between the bases.

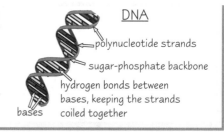

DNA

polynucleotide strands

sugar-phosphate backbone

hydrogen bonds between bases, keeping the strands coiled together

bases

Specific Base Pairing

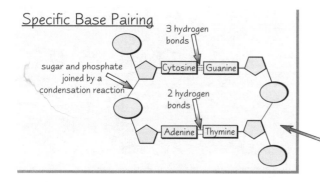

sugar and phosphate joined by a condensation reaction

3 hydrogen bonds

Cytosine ═ Guanine

2 hydrogen bonds

Adenine ─ Thymine

Each base can only join with one particular partner — this is called **specific base pairing**.

1)   In DNA **adenine** always pairs with **thymine** (**A - T**) and **guanine** always pairs with **cytosine** (**G - C**).

2)   It's the same in RNA, but **thymine**'s replaced by **uracil** (so it's **A - U** and **G - C**).

2 hydrogen bonds form between adenine and thymine.
3 hydrogen bonds form between guanine and cytosine.

DNA's Structure Makes it Good at its Job

1)   The job of DNA is to carry **genetic information**. A DNA molecule is very, very **long** and is **coiled** up very tightly, so a lot of genetic information can fit into a **small space** in the cell nucleus.

2)   Its **paired structure** means it can **copy itself** — this is called **self-replication** (see p.28). It's important for cell division and for passing on genetic information to the next generation.

The Genetic Code

DNA Contains the Basis of the Genetic Code

Genes are **sequences of nucleotides** in the DNA that code for specific **sequences of amino acids**.
The chains of amino acids make up particular **polypeptides** (proteins).
The way that DNA codes for proteins is called the **genetic code**.

1) Genes code for specific amino acids with sequences of three bases, called **base triplets**.
Different base triplets code for different amino acids.
For example AGA codes for serine and CAG codes for valine.

2) In the genetic code, each base triplet is read in sequence, separate from the triplet before it and after it.
Base triplets **don't share** their **bases** — so the code is described as **non-overlapping**.

3) The genetic code in DNA is also described as **degenerate**. This is because there are **more triplet codes** than there are amino acids. Some **amino acids** are coded for by **more than one base triplet**, e.g. Tyrosine can be coded for by TAT or TAC.

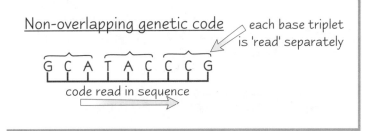

Practice Questions

Q1 What three things are nucleotides made from?

Q2 Which base pairs join together in a DNA molecule?

Q3 What type of bond joins the bases together?

Q4 What type of reaction joins nucleotides together to form a strand of DNA or RNA?

Q5 Which base is found in RNA but not in DNA?

Q6 What is a base triplet?

Q7 The genetic code is described as degenerate. What does this mean?

Exam Questions

Q1 Explain how the structure of DNA is related to its function. [2 marks]

Q2 Describe, using diagrams where appropriate, how nucleotides join together and
how two single strands of DNA become joined. [5 marks]

My genes are degenerate — there's a hole in the back pocket... (I'll get my code)

*You need to know the basic structure of DNA and RNA. Then there's the genetic code to get to grips with —
hmmm, rather you than me, but it **basically** comes down to the sequence of **bases**. I'm afraid there's nowt
else you can do except buckle down, pull your socks up and get all them facts learnt.*

DNA Replication

Here comes some truly essential stuff — DNA replication is the real nitty-gritty of biology. So eyes down for some serious fact-learning. I'm afraid it's all horribly complicated — all I can do is keep apologising. Sorry.

DNA can Copy Itself — Self-Replication

DNA has to be able to **copy itself** before **cell division** can take place —
this happens during the **interphase** period of the **cell cycle** (see page 34).
Cell division is essential for **growth and development** and **reproduction** — pretty important stuff.

1) **Specific base pairing** means that each type of base in DNA only pairs up with one other type of base — **A** with **T**, **C** with **G**. When a molecule of **DNA splits**, the **unpaired bases** on each strand can match up with complementary bases on **free-floating nucleotides** in the cytoplasm, making an **exact copy** of the DNA on the other strand. This happens with the help of the enzyme **DNA polymerase**. The result is **two molecules** of DNA **identical** to the **original molecule** of DNA:

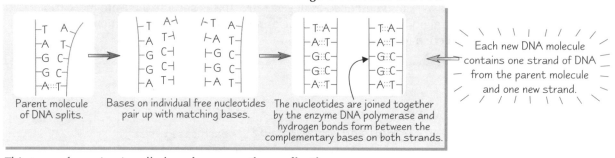

Parent molecule of DNA splits. | Bases on individual free nucleotides pair up with matching bases. | The nucleotides are joined together by the enzyme DNA polymerase and hydrogen bonds form between the complementary bases on both strands. | Each new DNA molecule contains one strand of DNA from the parent molecule and one new strand.

2) This type of copying is called **semi-conservative replication** — because **half** of each new molecule of DNA is from the **original** DNA molecule.

You can **Prove** that **Semi-Conservative Replication** Happens

You can prove that DNA replicates itself semi-conservatively by using two **isotopes** of **nitrogen** (DNA contains nitrogen) — heavy nitrogen (^{15}N) and light nitrogen (^{14}N).

1) Two samples of bacteria are grown — one in a nutrient broth containing **light** nitrogen, and one in a broth with **heavy** nitrogen. As the **bacteria reproduce**, they **take up nitrogen** from the broth to help make nucleotides for new DNA. So the nitrogen gradually becomes part of the bacteria's DNA.

2) A **sample of DNA** is taken from each batch of bacteria, and spun in a **centrifuge**. The DNA from the **heavy** nitrogen bacteria settles **lower** down the **centrifuge tube** than the DNA from the light nitrogen bacteria — because it's **heavier**.

3) Then the bacteria grown in the heavy nitrogen broth are **taken out** and put in a broth containing only **light nitrogen**. The bacteria are left for **one round of DNA replication**, and then **another DNA sample** is taken out and spun in the centrifuge.

4) This DNA sample settles out **between** where the light nitrogen DNA settled out and where the heavy nitrogen DNA settled out.

5) This means that the DNA in the sample contains a **mixture** of heavy and light nitrogen. The bacteria have **replicated semi-conservatively** in the light nitrogen. So the new bacterial DNA molecules contain **one strand** of the **old DNA** containing **heavy** nitrogen and **one strand** of **new DNA** containing **light** nitrogen.

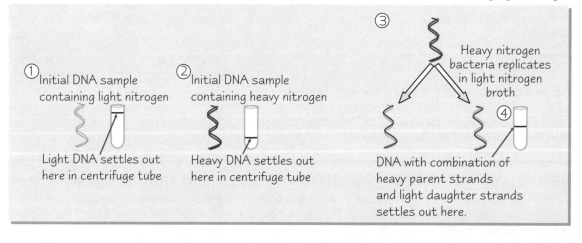

① Initial DNA sample containing light nitrogen — Light DNA settles out here in centrifuge tube

② Initial DNA sample containing heavy nitrogen — Heavy DNA settles out here in centrifuge tube

③ Heavy nitrogen bacteria replicates in light nitrogen broth

④ DNA with combination of heavy parent strands and light daughter strands settles out here.

Types of RNA

There are Three Types of RNA

There are **three types** of RNA, and all are involved in **making proteins** (see over the page for more on this — bet you can't wait).

Messenger RNA (mRNA)

1) **mRNA is a single polynucleotide strand** that's formed in the **nucleus**.

2) The important thing to know about it is that it's formed by using a section of a **single strand of DNA** as a template. **Specific base pairing** means that mRNA ends up being an exact **reverse copy** of the DNA template section (see the piccy on the right to make sense of this).

3) You also need to know that the **3 bases in mRNA** that pair up with a base triplet on the DNA strand are called a **codon**. Codons are dead important for making proteins (see p.30), so **remember this word**. Make sure you realise that a codon has the **opposite bases** to a base triplet (except the base **T** is replaced by **U** in **RNA**).

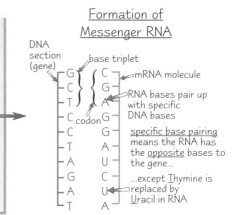

Formation of Messenger RNA

Transfer RNA (tRNA)

1) **tRNA is a single polynucleotide strand** that's folded into a **clover shaped molecule**.

2) Each tRNA molecule has a **binding site** at one end, where a specific **amino acid** attaches itself to the bases there.

3) Each tRNA molecule also has a specific sequence of **three bases** at one end of it, called an **anticodon**.

4) The significance of binding sites and anticodons are all revealed over the page. But you need to know **where they are found** on a tRNA molecule, so learn the diagram on the left off by heart.

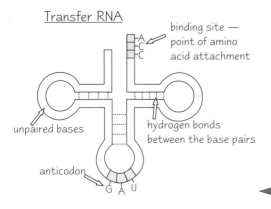

Transfer RNA

Ribosomal RNA (rRNA)

1) rRNA is made up of polynucleotide strands that are folded and attached to proteins to make things called **ribosomes** (see p.4).

2) Ribosomes are the site where proteins are made — and that's what the next page is all about...

Practice Questions

Q1 What is the name used to describe the type of replication in DNA?

Q2 What three types of RNA are there?

Q3 What is the name of the group of three bases on mRNA that correspond to a base triplet on DNA?

Q4 What shape does a chain of tRNA fold itself into?

Exam Question

Q1 Describe and explain the semi-conservative method of DNA replication. [6 marks]

Give me a D, give me an N, give me an A! What do you get? — very confused...

Quite a few terms to learn here — you're on the inescapable road to science geekville I'm afraid, and it's a road lined with crazy diagrams and strange words. DNA self-replication is sooo important — so make sure you understand what's going on. You need to learn the structure of the three types of RNA — it'll help you understand protein synthesis, on the next page.

Protein Synthesis

You've learnt about the genetic code and types of RNA — now you can learn all about their roles in making proteins. This stuff is biology at its most clever, and it's probably going on inside you right now. Weird.

Protein synthesis (making proteins) happens in **two stages** — transcription and translation. It involves both DNA and RNA.

First Stage — *Transcription* Occurs in the *Nucleus*

In **transcription** a '**negative copy**' of a **gene** is made. This copy is called **mRNA**.

Don't forget that, in RNA, adenine pairs up with uracil, not thymine.

1) A gene (a section of DNA) in the DNA molecule **uncoils** and the hydrogen bonds between the two strands in that section break, separating the strands.

2) One of the strands is then used as the **template** for transcription — it's called the '**sense strand**'.

3) Free **RNA nucleotides** in the nucleus line up alongside the template strand. Once the RNA nucleotides have **paired up** with their **complementary bases** on the DNA strand they're joined together by the enzyme **RNA polymerase**.

4) The strand that's formed is **mRNA**.

5) The mRNA moves out of the nucleus through a nuclear pore, and attaches to a **ribosome** in the cytoplasm, where the next stage of protein synthesis takes place.

6) When enough mRNA has been produced, the uncoiled strands of DNA re-form the hydrogen bonds and **coil back into a double helix**, unaltered.

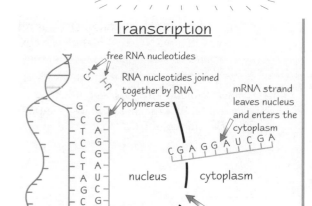

Transcription

Second Stage — *Translation* Occurs in a *Ribosome*

In **translation**, **amino acids** are stuck together to make a **protein**, following the order of amino acids coded for on the mRNA strand.

1) The **mRNA strand** has travelled to a ribosome in the cytoplasm, and attached itself.

2) All 20 **amino acids** needed to make human proteins are in the cytoplasm. tRNA molecules attach to the amino acids and transport them to the ribosome.

3) In the ribosome, a tRNA molecule binds to the start of the mRNA strand. This tRNA molecule has the **complementary anticodon** to the **first codon** on the mRNA strand, and attaches by **base pairing**. Then a second tRNA molecule attaches itself to the **next codon** on the mRNA strand in the **same way**.

4) The two amino acids attached to the tRNA molecules are joined together with a **peptide bond** (using ATP and an enzyme).

5) The first tRNA molecule then **moves away** from the ribosome, leaving its amino acid behind. The mRNA then **moves across** the ribosome by one codon and a third tRNA molecule binds to the **next codon** that enters the ribosome.

6) This process continues until there's a **stop codon** on the mRNA strand that doesn't code for any amino acid. You're left with a line of amino acids joined by peptide bonds. This is a **polypeptide chain** — the **primary structure** of a protein. The polypeptide chain moves away from the ribosome and translation is complete.

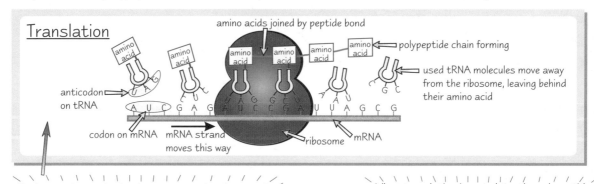

Translation

So, in protein synthesis, the sequence of codons on the mRNA strand determines the sequence of amino acids that makes up the primary structure of the protein.

When translation is complete, the polypeptide chain folds itself into its secondary and tertiary structure, and a protein is formed (see p.12).

Protein Synthesis

The **Structure of Enzymes** is **Determined** by **Protein Synthesis**

1) All **enzymes** are **proteins**, which are sequences of amino acids.

2) The amino acid sequences are determined by the base sequence in DNA, so **DNA** determines the **structure of enzymes**.

3) Enzymes speed up all our **metabolic pathways** (see p.16). They have a big influence over how our **genes** are **expressed physically**, by controlling chemical reactions required for growth and development. This physical expression of the gene contributes to the organism's **phenotype** (what the organism looks like).

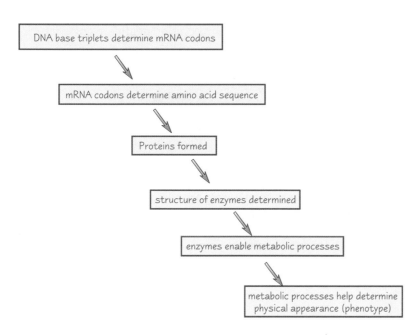

Kevin really knew how to express his jeans physically.

Practice Questions

Q1 What are the two main stages in protein synthesis?

Q2 Which base is not present in RNA?

Q3 Write the RNA sequence which would be complementary to the following DNA sequence. AATTGCGCCCG

Q4 Where does transcription take place?

Q5 Where does translation take place?

Q6 How is the DNA of an organism related to its phenotype?

Exam Questions

Q1 Explain the terms codon and anticodon. [2 marks]

Q2 Describe the process of protein synthesis. [10 marks]

mRNA codons join to tRNA anticodons?! — I need a translation please…

When you first go through protein synthesis it might make approximately no sense, but I promise its bark is worse than its bite. All those strange words disguise what is really quite a straightforward process — and the diagrams are dead handy for getting to grips with it. Keep drawing them yourself, 'til you can reproduce them perfectly.

Genetic Engineering

Genetic engineering is a dead popular exam topic because it shows how biology relates to real life — and examiners love all that stuff. These pages explain the process for manufacturing genes and, you've got to admit, it's kinda cool.

Recombinant DNA is like 'Home-Made' DNA

DNA has the **same** structure of **nucleotides** in **all organisms**. This means you can **join together** a piece of DNA from one organism and a piece of DNA from another organism.

DNA that has been **genetically engineered** to contain DNA from another organism is called **recombinant DNA**. It has **useful applications**. In the example below, a **human gene** coding for a **useful protein** is inserted into a bacterium's DNA. When the **bacterium reproduces**, the **gene** is **reproduced** too. The gene is **expressed** in the bacterium, and so the bacterium **produces** the **protein** which is coded for by the gene.

1) First you **find** the gene you want in the donor cell. The **DNA** containing the gene is **removed** from the cell and any **proteins** surrounding the extracted DNA are removed using **peptidase enzymes**.

2) Next you **cut out** the **useful gene** from the DNA using **restriction endonuclease** enzymes. These leave a **sticky end** (tail of unpaired bases) at each end of the useful gene. This stage is called **restriction**.

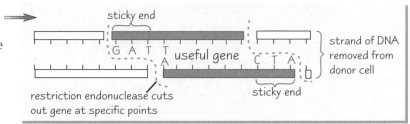

restriction endonuclease cuts out gene at specific points

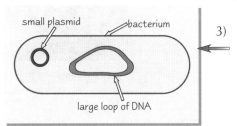

3) Then you need to **prepare** the other bit of DNA that you're **joining** the useful gene to (called a **vector**). The main vectors used are **plasmids** — these are small, **circular molecules** of DNA found in **bacteria**. They're useful as vectors because they can **replicate** without interfering with the bacterium's own DNA.

4) Finally, you need to **join** the **useful gene** to the **plasmid vector DNA** — this process is called **ligation**. This is where the **sticky ends** come in — **hydrogen bonds** form between the **complementary bases** of the sticky ends. The DNA molecules are then 'tied' together with the enzyme, **ligase**. The new DNA is called **recombinant DNA**.

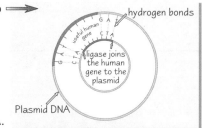

The useful gene is now in the recombinant DNA of a plasmid vector in a bacterium. So now the bacterium is made to reproduce loads, so you get loads of copies of the gene (see next page for how this happens). Clever, huh.

Reverse Transcriptase makes DNA from mRNA

Transcription is when mRNA is made from DNA (see p.30). **Reverse transcriptase** is a nifty enzyme that runs this process backwards and makes DNA from mRNA. It's useful in genetic engineering because **cells** that make **specific proteins** usually contain **more mRNA molecules than genes**. For example, **pancreatic cells** produce the protein **insulin**. They have loads of mRNA molecules on their way to ribosomes to make the insulin — but only **two copies** of the DNA gene for insulin.

1) **mRNA** is **extracted** from donor cells.

2) The mRNA is mixed with **free DNA nucleotides** and **reverse transcriptase**. The reverse transcriptase uses the mRNA as a **template** to synthesise a new strand of **complementary DNA**.

3) The complementary DNA can be made **double-stranded** by mixing it with DNA nucleotides and **polymerase enzymes**. Then the useful gene from the double-stranded DNA is inserted into a plasmid (see section above) so the bacteria can make lots of the product of the gene.

4) The diagram shows a monkey swinging from a strand of DNA. Try to ignore him — he's attention-seeking.

Genetic Engineering

Genetic Engineering has *Important Medical Uses*

Sometimes humans can't produce a certain **protein** because the **gene** that codes for it is **faulty**.
By creating recombinant DNA, genes can be artificially **manufactured**, to replace the faulty genes.
This has important **medical benefits**.

For example:

1) **Diabetics** don't have the healthy gene needed to make the protein **insulin**, which controls blood sugar levels.
2) **Haemophiliacs** lack the healthy gene that codes for the protein **factor VIII**, which allows blood to clot.
3) Genetic engineering allows these genes to be manufactured in bacterial cells and used to help sufferers.

Industrial Fermenters are Used for Large Scale *Protein Production*

Once a useful gene is in a bacterium, it has to start working to produce the **protein**. An **industrial fermenter** is used to culture the bacteria and produce a **large amount** of the gene product (the protein).

1) A **promoter** gene is often included along with the **donor** gene when the **recombinant DNA** is made. This 'switches on' the useful donor gene so it starts making the **protein**.
2) The **bacteria** containing the recombinant DNA are grown or cultured in a **fermenter**. Inside the fermenter, they're given the **ideal conditions** needed for rapid growth. They **reproduce quickly**, until there are millions of bacteria inside the fermenter.
3) The **plasmids**, including plasmids made of recombinant DNA, **replicate** at each cell division — so each new bacterium contains the useful gene.
4) As the bacteria grow, they start producing the **human protein** that the donor gene in the plasmid codes for, e.g. **human factor VIII** or **human insulin**.
5) The bacteria can't use human protein, so it **builds up** in the medium inside the fermenter. When enough has built up, it can be **extracted** and processed for use.

Practice Questions

Q1 What is recombinant DNA?

Q2 What is a vector?

Q3 Name one type of vector commonly used in genetic engineering.

Q4 What does reverse transcriptase do?

Q5 What proteins do diabetics and haemophiliacs lack?

Q6 Where are the bacteria containing recombinant DNA usually grown?

Exam Question

Q1 Describe how genetic engineering is used to help people with haemophilia. [6 marks]

Monkey vectors — Plasmid of the Apes...

You see, biology isn't just an evil conspiracy to keep students busy — it has loads of really important uses in real life. For example, sufferers of genetic diseases now have a far greater chance of having successful treatment, which is nice. Round of applause for biology, that's what I say.

The Cell Cycle and Mitosis

I don't like cell division. There, I've said it. It's unfair of me, because if it wasn't for cell division I'd still only be one cell big. It's all those diagrams that look like worms nailed to bits of string that put me off.

Mitosis is the Cell Division Used in Asexual Reproduction

Cells increase in number by **cell division**. There are two types of cell division — **mitosis** and **meiosis**. **Mitosis** produces daughter cells that are **genetically identical** to the parent cell. It's needed for the **growth** of multicellular organisms (like us) and for **repairing** damaged tissues. It's also the means of **asexual reproduction**. **Asexual** reproduction needs only **one** parent. The offspring are normally **genetically identical** to this parent and are called **clones**. The dandelion plant is an example of an organism which reproduces asexually.

Cells from multicellular organisms have a clear **cell cycle** that starts when they are produced by cell division, and ends with them dividing themselves to produce more identical cells. The cell cycle consists of a period of cell division (**mitosis**) and a period in between divisions called **interphase**. Interphase is subdivided into 3 separate growth stages. These are called **G₁**, **S** and **G₂**. Each stage involves specific cell activities:

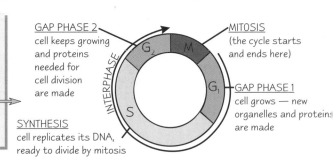

GAP PHASE 2 — cell keeps growing and proteins needed for cell division are made

MITOSIS (the cycle starts and ends here)

GAP PHASE 1 — cell grows — new organelles and proteins are made

SYNTHESIS — cell replicates its DNA, ready to divide by mitosis

Mitosis has Four Main Stages

Mitosis is really one **continuous process**, but it's described as a series of **division stages** — **prophase**, **metaphase**, **anaphase** and **telophase**. **Interphase** comes **before** the division stages — so that cells can grow and prepare to divide by replicating their DNA.

1) **Prophase** — The chromosomes **condense**, getting shorter and fatter. Tiny bundles of protein called **centrioles** start moving to opposite ends of the cell, forming a network of protein fibres across it called the **spindle**. The nuclear envelope (membrane) breaks down and chromosomes lie free in the cytoplasm.

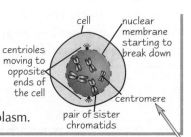

cell
nuclear membrane starting to break down
centrioles moving to opposite ends of the cell
centromere
pair of sister chromatids

So long and thanks for all the organelles!

It's so hard letting go of my baby girls. It feels like a part of me has gone with them.

There, there, love — it's all part of the cycle of life.

Mitosis can be a moving time.

2) **Metaphase** — The chromosomes (each with two chromatids) line up along the middle of the cell and become attached to the spindle by their centromere.

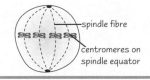

spindle fibre
centromeres on spindle equator

As mitosis begins, the chromosomes are made of two strands joined in the middle by a <u>centromere</u>. The separate strands are called <u>chromatids</u>. There are two strands because each chromosome has already made an <u>identical copy</u> of itself during <u>interphase</u>. When mitosis is over, the chromatids end up as one-strand chromosomes in the new daughter cells.

3) **Anaphase** — The centromeres attaching the chromatids to the spindles divide, separating each pair of sister chromatids. The spindles contract, pulling chromatids to opposite poles, centromere first.

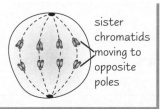

sister chromatids moving to opposite poles

4) **Telophase** — The chromatids reach **opposite poles** of the cell on the spindle. They uncoil and become long and thin again. They're now called **chromosomes** again. A **nuclear envelope** (membrane) forms around each group of chromosomes, so there are now **two nuclei**. The **cytoplasm divides** and the cell membrane pinches inwards to give **two daughter cells** which are **identical** to the original cell. Mitosis is finished and each daughter cell starts the **interphase** part of the cell cycle to get ready for the next round of mitosis.

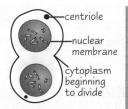

centriole
nuclear membrane
cytoplasm beginning to divide

The Cell Cycle and Mitosis

The Daughter Cells Are Genetically Identical to the Parent Cell

Excuse me if I keep going on about this, but it's the most important thing to remember about mitosis. The **chromosome number** of the new nuclei is **the same** as that of the original nucleus, and the new cells contain exactly the **same genetic information** as each other. That's why mitosis is so useful for **replacing** cells — for example, in the human body some of the cells that divide by mitosis most often are those that line the **gut**, because these cells are constantly worn away by the passage of food.

Cancer is the Result of Uncontrolled Cell Division

Cell division is controlled by genes. Normally, when a cell has divided enough times to make **enough new cells**, it stops. But if there's a **mutation** in the gene that controls cell division, the cells **grow out of control**. (Mutations are changes in the base sequence of an organism's **DNA**.) The cells **keep on dividing** to make more and more cells — which form a **tumour**. Tumours can be broken up and **carried** around the body in **blood**, getting lodged in different parts of the body. They keep on **growing uncontrollably**, squashing the normal tissues around them.

Genes that have mutated to cause cancer are called <u>oncogenes</u>.

Mutations happen naturally by chance. But there are some substances, called **carcinogens**, that increase the chance of a normal cell becoming a cancer cell — like **tobacco smoke**, **radiation**, **ultraviolet light** from the **Sun** and **x-rays**.

Practice Questions

Q1 What is meant by asexual reproduction?

Q2 What is cell division by mitosis used for?

Q3 Explain why the chromosome becomes visible as two chromatids at the end of prophase.

Q4 Explain how a gene mutation can cause cancer.

Q5 Give two examples of carcinogens.

Exam Questions

Q1 The diagrams show cells at different stages of mitosis.

Cell A   Cell B   Cell C

a) For each of the cells A, B and C state the stage of mitosis, giving a reason for your answer. [6 marks]
b) Name the structures labelled X, Y and Z in cell A. [3 marks]

Q2 During which stages of the cell cycle would the following events take place?
a) DNA replication. [2 marks]
b) Formation of spindle fibres. [2 marks]

Doctor, I'm getting short and fat — don't worry, it's just a phase...

Quite a lot to learn in this topic — but it's all dead important stuff so no slacking. Mitosis is vital — it's how cells multiply and how organisms like us grow and develop. Remember that chromosomes are in fact usually made up of two sister chromatids joined by a centromere. Aaw, nice to know family values are important to genetic material too.

Meiosis and Sexual Reproduction

More cell division — lovely jubbly. Meiosis is the cell division used in sexual reproduction.
It consists of two divisions, not one, and it's used to produce ova and teeny sperms.

DNA is Stored as Chromosomes in the Nuclei of Cells

Chromosomes are thread-like structures made up of long, tightly coiled molecules of DNA.

A **karyotype** is a **drawing** made from a photo of all the **chromosomes** in a cell. All the chromosomes are matched and put into **pairs** in order of decreasing size. Pairs of matching chromosomes are called **homologous chromosomes**. In a homologous pair both chromosomes are the same shape and size and have the **same genes** in the same location. One chromosome in each homologous pair comes from the female parent, and the other from the male parent.

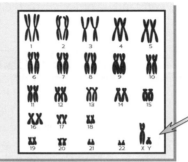

Human cells have 23 pairs of chromosomes, but only 22 pairs are homologous — the 23rd pair consists of the sex chromosomes (X and Y in men, and X and X in women).

In pictures like this, each chromosome you see is really a **double structure**. The DNA has replicated to give **two** genetically identical DNA molecules, which are tightly grouped together. These are called **sister chromatids**.

Human cells contain 46 single chromosomes in total — this is called the **diploid number** (**2n**). Different species have different diploid numbers. **Gametes** (sex cells) contain **half** the diploid number of chromosomes — this is called the **haploid number** (**n**).

DNA From One Generation is Passed to the Next by Gametes

1) **Gametes** are the **sperm** cells in males and the **ova** (egg cells) in females. They join together at **fertilisation** to form a **zygote**, which divides and develops into a **new organism**.

2) Normal **body cells** have the **diploid number** (**2n**) of chromosomes — meaning each cell contains **two** of each chromosome, one from the mum and one from the dad.

3) But **gametes** have a **haploid** (**n**) number of chromosomes — so there's only one of each chromosome.

4) At **fertilisation**, a **haploid sperm** fuses with a **haploid egg**, making a cell with the normal diploid number of chromosomes. Half these chromosomes are from the **father** (the sperm) and half are from the **mother** (the egg).

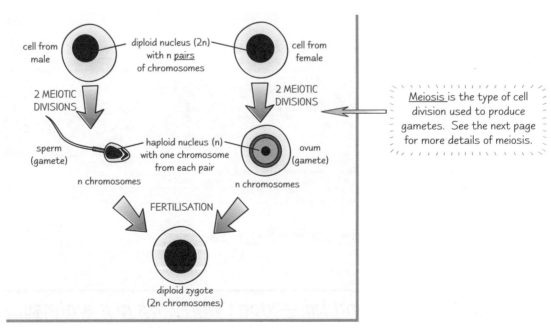

Meiosis is the type of cell division used to produce gametes. See the next page for more details of meiosis.

Meiosis and Sexual Reproduction

Meiosis Halves the Chromosome Number

1) Cells that divide by meiosis are **diploid** to start with, but the cells that result from meiosis are **haploid**. Without meiosis, you'd get **double** the number of chromosomes in each generation, when the gametes fused.

2) Meiosis happens in the **reproductive organs**. In humans it's in the **testes** for males and the **ovaries** for females. In plants it's in the **anthers** and **ovules**.

3) Unlike mitosis, there are **two divisions**. This **halves** the chromosome number.

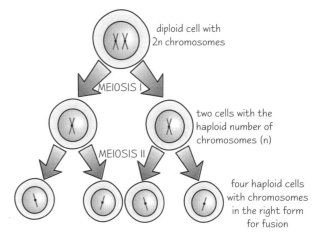

diploid cell with 2n chromosomes

MEIOSIS I

two cells with the haploid number of chromosomes (n)

MEIOSIS II

four haploid cells with chromosomes in the right form for fusion

The two divisions in meiosis are called **meiosis I** and **II**.

1) In **Meiosis I** the **homologous pairs** of **chromosomes** are separated, which **halves** the number of chromosomes in the daughter cells.

2) **Meiosis II** is like mitosis — it separates the **pairs of chromatids** that make up each chromosome.

3) Unlike mitosis, which results in two genetically identical diploid cells, meiosis results in **four haploid cells** (gametes) that **are genetically different** from each other.

Practice Questions

Q1 Explain what is meant by the terms "haploid" and "diploid."

Q2 What is a karyotype?

Q3 How many divisions are there in meiosis?

Q4 In which organs in a human would meiosis take place?

Exam Question

Q1 The diagram shows stages of meiosis in a human ovary. Each circle represents a cell.

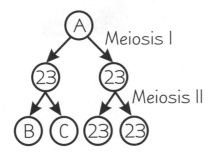

A

Meiosis I

23   23

Meiosis II

B C 23 23

a) How many chromosomes would be found in cells A, B and C? [3 marks]

b) Explain why it's important for gametes to have half the number of chromosomes as normal body cells. [2 marks]

Reproduction isn't as exciting as some people would have you believe...

*For some reason, this stuff can take a while to go in (insert own joke). But that's no excuse to just sit there staring frantically at the page and muttering "I don't get it," over and over again. Use the diagrams to help you understand — they look evil, but they really help. The key thing is to understand what happens to the **number of chromosomes** in meiosis.*

Energy Transfer Through an Ecosystem

Two nice, easy pages — loads of it's common sense and some stuff might already be familiar.
Just be careful when you're learning all the definitions, because some of the words mean very similar things.

There are **Ten Main Definitions** *You* **Need To Know**

TERM	MEANING
Ecosystem	An ecosystem supports life. Nutrients are recycled though an ecosystem, and energy flows through an ecosystem. E.g. a pond, a lawn, a wood. An ecosystem includes both the living and the non-living things there.
Habitat	A place where an organism lives, e.g. a wood pigeon's habitat is a wood.
Community	**All** the organisms living in a particular ecosystem, e.g. a woodland community (which would include all the animals, plants etc. that live there.)
Population	A group of organisms of the **same species** living in the same place at the same time, e.g. a **population** of wood pigeons living in a wood, or a **population** of woodlice living under a log.
Niche	The role played by an organism in a community (like its behaviour, what it eats and what eats it). Each species has its own unique niche.
Producer	Producers make their own food using an external energy source, e.g. light from the Sun. Plants are producers. Organisms that make their own food are also called **autotrophs**.
Consumer	Consumers eat other organisms for food and energy. Cows eat a **producer** (grass) so they're **primary** consumers. Humans then eat cows so they're **secondary** consumers.
Trophic level	A particular **feeding stage** in the food chain, e.g. producer, primary consumer, secondary consumer, etc.
Food chain	A sequence representing the way energy flows from one organism to another.
Food web	A diagram showing all the feeding relationships between the organisms of a **community**. It's made up of many inter-connected food chains.

Energy Flows Through *Food Chains*

1) Each organism in a food chain is at a different **trophic level** — it can be a producer, a primary consumer, a secondary consumer, a tertiary consumer, etc.

2) Food chains start with a **producer**.

3) The **primary consumer** eats the producer, and so on up the food chain.

Example of a Food chain:

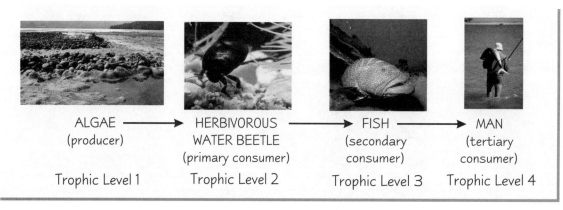

ALGAE ———→ HERBIVOROUS ———→ FISH ———→ MAN
(producer)   WATER BEETLE   (secondary   (tertiary
                 (primary consumer)   consumer)   consumer)

Trophic Level 1   Trophic Level 2   Trophic Level 3   Trophic Level 4

Transfer of energy between trophic levels isn't very efficient. Only about **10%** of the **energy** stored in the organisms at one trophic level passes to the organisms in the next level. The energy is lost in a few different ways:

1) Some is lost as **heat** from the organisms.
2) Some is lost in their **waste**.
3) Energy is used by organisms during respiration, to let them move, grow, reproduce etc.
4) Animals don't usually eat **all** of the food available in the organism they feed on, and can't digest everything they do eat.

Energy Transfer Through an Ecosystem

Food Webs are Made Up of Many Inter-Linked Food Chains

Most organisms don't eat just one thing — so food chains don't tell the full story.
Food webs tell you more — for example, this one shows that algae are eaten by
three different primary consumers, not just by the water beetle. Like food
chains, food webs always start with producers and end with the top consumers.

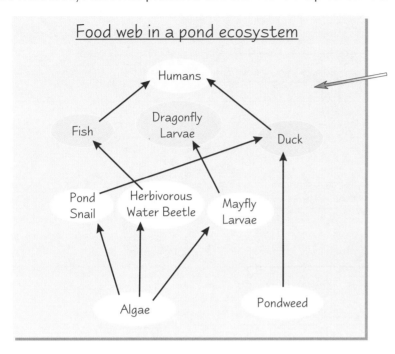

It's not always easy to understand how
humans made it to the top of the food chain.

Practice Questions

Q1   What is an ecosystem?

Q2   What is meant by the term "niche"?

Q3   What is the difference between a community and a population?

Q4   At which trophic level would you find a producer?

Q5   Give an example of a secondary consumer.

Q6   Name three ways that energy is lost as it is transferred between trophic levels.

Q7   Why are food webs more useful than food chains?

'Arbitrary' just means that the
units could be anything. It
doesn't matter what they are,
so you don't have to worry
about them — which is nice.

Exam Question

Q1   Approximately 1% of the sun's energy is incorporated into the producers. Only 10% of the energy in one
trophic level passes to the organisms in the next trophic level. If 10,000 arbitrary units of energy are given
out by the sun, how many of these arbitrary units will be incorporated into the secondary consumers?   [3 marks]

Things eat each other — how do they make it so complicated?

*A good question, but not one you've got time to ponder now. What you need to do is learn that list of 10 definitions.
Some are fairly obvious, like producer, but some are more tricky, like the difference between a population and a
community. Don't try to understand the reasons behind the terms, it's a waste of energy. Just get them learnt.*

The Nitrogen Cycle

At last — the page we've all been waiting for. It's every biology student's favourite friend, the nitrogen cycle. It may not look very pleasant, but it's really not that bad once you get to know it.

The **Nitrogen Cycle Diagram** is Scary but **Important**

OK, here goes. All those arrows and boxes look alarming, but actually it's pretty straightforward, once you understand the processes that are going on. Link it up to the explanation on the next page and it'll soon start to make sense.

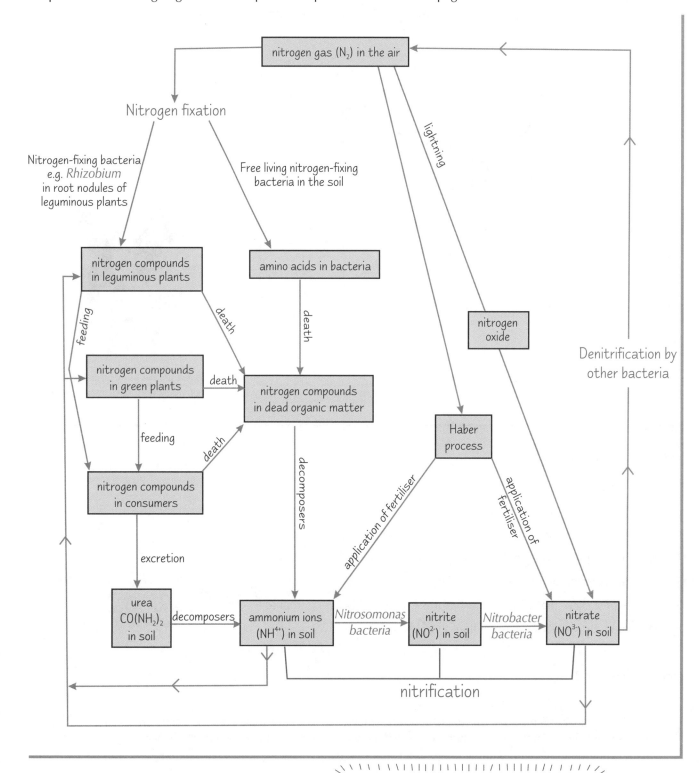

It's easy to get nitrogen fixation muddled up with nitrification and denitrification. Make sure you're confident about which is which, and about the bacteria that belong with each process.

The Nitrogen Cycle

Micro-organisms are Important in the Nitrogen Cycle

Don't let the diagram on the last page scare you — it won't take as long as you think to learn it. Honest.
Basically, **all living things** need **nitrogen** to make molecules like **proteins**, **ATP** and **DNA**. Plants can only absorb nitrogen in the form of **ammonium** or **nitrate ions**, while **consumers** get their nitrogen by eating other organisms.

Here's a list of the facts to help you understand the diagram:

1) Some of the **ammonium** and **nitrate ions** used by plants are in the soil because of **nitrogen fixation**. This is when **atmospheric nitrogen** is converted to ammonium or nitrate ions by **nitrogen fixing bacteria**. Some nitrogen fixing bacteria can live freely in the soil. Others, like *Rhizobium*, live **symbiotically** in the root nodules of leguminous plants such as peas and beans. In a symbiotic relationship, both organisms benefit — in this example the bacteria get sugars from the plant, and the plant gets a supply of nitrogen it can use.

2) Energy from **lightning** also helps turn atmospheric nitrogen into nitrates.

3) The industrial **Haber process** uses atmospheric nitrogen to make nitrate and ammonia fertilisers. Using these fertilisers increases the concentration of ammonium and nitrate ions in the soil.

4) **Decomposers** turn nitrogen-containing compounds in **dead organisms** into **ammonium ions**. These are then converted to **nitrite** by *Nitrosomonas* bacteria, and the nitrite's converted to **nitrate** by *Nitrobacter* bacteria. The ammonium to nitrate process is called **nitrification**.

5) **Denitrification** is when nitrates are turned back into atmospheric nitrogen by bacteria. This only happens in the absence of oxygen — e.g. in waterlogged soil (because these denitrifying bacteria need **anaerobic** conditions).

Practice Questions

Q1 In what form do plants absorb nitrogen from the soil?

Q2 What role does lightning play in the nitrogen cycle?

Q3 Explain the symbiotic relationship which is important in the nitrogen cycle.

Exam Question

Q1 Give an account of the nitrogen cycle, highlighting the importance of microbes. [10 marks]

Look at the state of that nitrogen cycle — I want my mum...

It looks scary, I won't deny it. But once you start it's really not hard to learn. The best way to learn the cycle is to try and draw the diagram (no peeking), see which bits you missed, then try again. When you've managed it perfectly three times in a row it's probably in your head for good, but come back to it now and again just to check.

Health and Disease

Health is more than just the absence of disease. The World Health Organisation defines health as "a state of complete physical, mental and social well being". But you still need to learn about diseases, though — bummer.

There are **Different Categories** of **Disease**

You need to learn all these categories of disease, and an example for each one.

NB — a disease can fall into more than one of these categories. For example, you could have a self-inflicted physical disease or a non-infectious degenerative disease.

Type of Disease	Description	Example
Physical	Temporary or permanent damage to your body.	torn knee cartilage
Mental	Psychological disorder.	depression
Social	Disease caused by social conditions or environment.	obesity or asbestos poisoning
Infectious	Caused by another organism which enters the body.	influenza virus
Non-infectious	Can't be passed from one person to another.	cancer
Degenerative	Age-related wear and tear of tissues or organs.	osteo-arthritis
Inherited	Defective genes are passed on from parents.	cystic fibrosis
Self-inflicted	Damage you have done to yourself.	emphysema from smoking
Deficiency	A health problem caused by inadequate diet.	scurvy due to lack of vitamin C

Infectious Diseases are Caused by **Micro-Organisms**

There are many different **micro-organisms** that can cause infectious disease — they can be viruses, bacteria, fungi, algae or protozoa.

1) Organisms that cause infectious diseases are called **pathogens**.
2) **Animal diseases** are usually caused by bacteria or viruses.
3) **Plant diseases** are usually caused by fungi or viruses.

BUT, it's important to remember that not all micro-organisms are harmful.

Statistics can Reveal the **Health** of a **Population**

Epidemiology is the **study of patterns of disease** using statistics. Most epidemiological studies look at either **incidence** of disease (the **number of cases** over a period of time) or **mortality** (the **number of deaths** from a disease over time).

1) Epidemiology is often used to **identify links** between a disease and its **cause**. For example in the 1950s, epidemiological studies linked smoking to deaths from lung cancer. More recently, studies have linked sunbathing with skin cancer.

2) Epidemiology is also used to decide **suitable cures** and treatments for diseases:

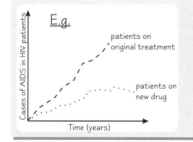

This graph shows that fewer HIV patients developed AIDS when given a new drug. This would go some way towards proving that the new drug is a suitable treatment for HIV patients.

Diseases can **Spread Across** the **World**

You need to know these terms about how diseases **spread**.

1) **Pandemic** — the spread of disease **internationally**, e.g. SARS or HIV.
2) **Epidemic** — the **rapid spread of disease** through a population, e.g. new strains of 'flu.
3) **Endemic** — a common disease, **always present** in a population, e.g. measles.

Health and Disease

Standards of Health Care Vary Between Countries

Wealth **varies** between countries — the world is often described as being made up of **More Economically Developed Countries (MEDCs) and Less Economically Developed Countries (LEDCs)**. Health care is generally worse in LEDCs — so their populations suffer more from diseases.

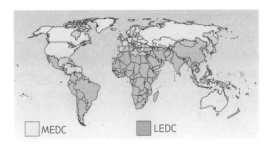

MEDC LEDC

Generally in **LEDCs**:

1) People suffer **more deaths from infectious diseases** (like malaria and TB).

2) People have a **lower life expectancy**.

3) There's poor medical care, housing and diet, which **increases the incidence of disease**.

4) There are **high infant mortality rates** from diseases like diarrhoea and measles.

Generally in **MEDCs**:

1) People suffer **fewer deaths from infectious diseases**.

2) People have a **higher life expectancy**.

3) There are better medical services, housing and diet, which **lowers the incidence of disease**.

4) Children are immunised against infectious diseases early on, so there are **low infant mortality rates**.

5) People suffer **more deaths from degenerative diseases** due to unhealthy lifestyles — things like coronary heart disease, obesity and cirrhosis of the liver due to alcoholism.

Results from the Human Genome Project May Help Treat Disease

The **Human Genome Project** is an international project that has **identified** every human gene and now aims to work out what each one **codes for**. This has **potential advantages** for improving health:

1) We might be able to change our **genetic inheritance** by eliminating genes that code for inherited diseases.

2) Doctors can make more **accurate diagnoses** of diseases.

3) Better drugs can be made to **target** specific diseases.

4) Drugs with **fewer side-effects** can be made.

5) There'll be more information to base **medical research** on.

Many people have ethical arguments against changing our natural genetic make-up. Some scientists are attempting to clone humans and it could be possible in the future to choose a 'designer' baby, with the genetic make-up of your choice. Many people dislike the idea of this level of interference.

Practice Questions

Q1 Define health in relation to disease.

Q2 Give an example of a deficiency disease.

Q3 What can epidemiological studies tell us?

Q4 What's the difference between endemic and pandemic diseases?

Q5 What's the aim of the Human Genome Project?

Exam Questions

Q1 Discuss the advantages and disadvantages of the Human Genome Project in relation to the treatment of diseases. [4 marks]

Q2 Over 12 million children in Africa and Asia under the age of 5 years die from diseases every year. Give reasons for the spread of disease in these continents. [4 marks]

The Human Gnome Project — it started in my garden...

Good mix of facts to learn and ethical issues to get your teeth into here. If the ethical side of manipulating genes to prevent disease crops up in the exam you need to be able to show that you're aware of both sides of the arguments. To clone or not to clone — that is the (essay) question.

A Balanced Diet and Essential Nutrients

Two pages all about diet — no glossy mag would be complete without them.
Now here comes the science part — concentrate. Why? Because you're worth it.

A *Balanced Diet* Supplies All the *Essential Nutrients*

A balanced diet gives you all the **essential nutrients** you need, plus fibre and water.
The 5 essential nutrients are **carbohydrates**, **proteins**, **lipids**, **vitamins** and **minerals**.

Large molecules are broken down into small molecules during digestion.
Nutrients are absorbed as small molecules through **microvilli** in the small intestine.

1) **Carbohydrates** are broken down and absorbed as **monosaccharides**.

2) **Lipids** are broken down and absorbed as **fatty acids** and **glycerol**.

3) **Proteins** are broken down and absorbed as **amino acids**.

Different nutrients are required for different functions in the body.

NUTRIENTS	FUNCTIONS
Carbohydrates	Provide energy.
Lipids	Store energy, provide insulation, constituent of cell membranes, physically protect organs.
Proteins	Growth and repair of tissue.
Vitamins	Various functions: e.g. vitamin B is needed for ATP production; vitamin K is needed for blood clotting.
Mineral ions	Various functions: e.g. iron is needed for healthy blood; calcium is needed for bone formation and strength.
Fibre	Keeps the gut in good working order.
Water	Used in hydrolytic reactions. We need a constant supply to replace water lost through urinating, breathing and sweating.

Energy and *Nutrient* Needs *Vary* Between People

Different factors affect our energy and nutrient needs:

Age → **Young people** need more protein for growth. **Older people** need more **calcium** to protect against **degenerative bone diseases** like **osteo-arthritis**.

Gender → **Females** need more **iron during menstruation** due to loss of blood.

Pregnancy → Pregnant women need **more vitamin A** — it's essential for **bone formation** in the **foetus**.

Lactation → **Breast-feeding mothers** need extra protein, vitamin D and iron.

Physical Activity → **Active people** need more **protein** for muscle development and more **carbohydrate** for energy.

DRVs are *Dietary Reference Values*

DRVs are **recommendations** made by the Government on estimated **energy and nutritional requirements** for particular groups of the UK population. The recommendations are related to age, gender, level of fitness and body size.

These are all the codes used in working out DRVs, so you need to learn 'em.

DRV   **Dietary Reference Values** — Government recommendations based on scientific studies.

EAR   **Estimated Average Requirement** — of energy or nutrients.

BMR   **Basal Metabolic Rate** — the speed at which energy is used up in the body.

PAL   **Physical Activity Levels** — varies with type and length of activity and your own body size.

DRVs should be used when assessing or planning the dietary intake of a group. People have **different energy needs** depending on the rate at which energy is used up in their bodies. If you burn energy quickly then you have a **high BMR**. Physical activity **increases BMR**. Calculations on energy needs can be made using DRVs.

The equation used is **EAR = BMR x PAL**

A Balanced Diet and Essential Nutrients

Essential Amino Acids and Fatty Acids have to be Obtained from Food

1) There are **10** amino acids that the body can't make. These are called **essential amino acids** because it's essential that they're in our food.

2) Some essential amino acids **make non-essential ones**. For example, the essential amino acid **phenylalanine** can be converted into the non-essential amino acid **tyrosine**.

3) Similarly, our body makes some fatty acids, but **essential fatty acids** can't be made by cells — they have to be provided by our diet.

> Essential amino acids and fatty acids have many uses —
> 1) Amino acids are used to **make proteins** required by the body.
> 2) Essential fatty acids **heal wounds** and prevent hair loss.
> 3) Fatty acids also ensure healthy growth in **babies**.
> 4) **Linolenic acid** is an essential fatty acid that makes the phospholipids found in cell membranes and lowers cholesterol levels in blood plasma. It can also be converted into lots of other fatty acids.

The Body Needs Small Amounts of Vitamins for Specific Functions

Vitamins are a group of chemicals needed to keep you healthy. They're only required in very **small amounts**. You need to know the **functions** of vitamins **A** and **D**:

1) You get **vitamin A** from **meat** and especially from **vegetables**. It's essential for keeping **epithelial cells** in good condition and for **healthy eyesight**.

2) **Vitamin D** is found in oily fish and eggs. It **controls calcium absorption** to make strong bones and teeth. It also helps **absorb the phosphorous** needed for making **ATP** and **nucleic acids**.

Practice Questions

Q1 Name the uses of carbohydrate and protein in the body.

Q2 Why would someone with a sedentary lifestyle need to eat less food than a professional athlete?

Q3 What use does water have in the body?

Q4 What does BMR stand for?

Q5 What is an essential amino acid?

Q6 Which vitamin:
a) is needed for strong teeth?
b) improves vision?

Exam Questions

Q1 What is the estimated average requirement of energy for a 16 year old female with a low physical activity level of 1.4 and a basal metabolic rate of 7.2 MJ per day? [3 marks]

Q2 Describe why humans need essential amino acids. [4 marks]

I have a balanced diet — one doughnut for every kebab...

Everyone needs the same essential nutrients, but how much we need of each depends on things like age, job and gender — champion cheese-hurlers will need more carbohydrates than champion chess players. As you're learning all the posh names (like phenylalanine), learn the spelling too. It'll save time thinking about it in the exam and will help you remember them.

Malnutrition

Two rather sobering pages for you. We need food to survive — simple as that. The sad thing is, loads of people in the world don't have enough to eat, whilst other people suffer from problems due to over-eating. It's not a fair world.

Malnutrition is When You Don't Get the Proper Nutrients

Malnutrition is basically having too little or too much of some nutrient. There are four causes:

1) Not having **enough food**. This can lead to **starvation**.

2) Having an **unbalanced diet** — so you aren't getting all the essential nutrients you need from your diet. This leads to all kinds of **deficiency illnesses**.

3) **Defective assimilation** of nutrients — this means your body isn't absorbing the nutrients from digestion into your bloodstream properly. This also causes **deficiency illnesses**.

4) Having **too much** food. This can lead to **obesity**.

Protein-energy Malnutrition (PEM) — not getting enough Protein and Energy

Inadequate **protein** intake means **growth** and **development** slow down. Inadequate **energy** causes the **basal metabolic rate** to slow down so that physical activity becomes difficult.

1) **Children** are particularly vulnerable to PEM because they need extra protein and energy to help them grow. Children suffering from PEM have **muscle wasting**, **skin rashes** and **swollen abdomens**.

2) PEM can cause **failure of the pancreas** and **intestines**.

3) Extreme forms of PEM lead to the deficiency diseases **marasmus** and **kwashiorkor**. These are more common in **LEDCs**.

4) If it carries on long enough, a child's growth and development stops completely, causing permanent damage.

Vitamin D Deficiency Affects Healthy Bone Development

Vitamin D is involved in **bone formation**. It stimulates **gut cells** to absorb calcium, and stimulates **bone cells** to use the stored calcium to form bones. So if you don't get enough vitamin D in your diet, your bones don't form properly and you can get one of two deficiency diseases:

1) **Rickets** — bone softening in children, which causes legs to become bowed. It also causes enlargement of the liver and the spleen.

2) **Osteomalacia** — the same bone softening symptoms, but in adults.

Vitamin A Deficiency Affects your Eyes and Immune System

You get vitamin A from meat and vegetables. It's used for two main things in the body:

1) **Epithelial cells** (in the digestive tract, skin and lungs) convert it to **rhetonoic acid** — used to **fight pathogens**.

2) **Rod cells** in the eye convert it into **rhodopsin**, which is used to help us **see in dim light**.

Deficiency in Vitamin A can cause three health problems:
- **poor defence** against pathogens,
- **poor night vision** or
- **xerophthalmia**, where the skin of the cornea is scarred, causing blindness.

Malnutrition

Over-Nutrition and Lack of Exercise can Lead to Obesity

Some people, especially in **MEDCs**, have **dietary diseases** caused by **too much food**.

1) **Obesity** is a common dietary disease — it's defined as being **20% (or more) over recommended body weight**.

2) **Too much sugary or fatty food** and **too little exercise** are the main causes of obesity.

3) People can also be obese due to an **underactive thyroid gland**, but this problem isn't common.

4) Obesity can increase the risk of **diabetes**, **arthritis**, **high blood pressure, coronary heart disease** (CHD) and even some forms of **cancer**.

I'm just big boned.

Anorexia Nervosa is a Particular Form of Starvation

Anorexia nervosa is a **mental disease** with **physical effects**. It's caused by an **emotional refusal to eat**.

1) Anorexia is usually caused by **low self esteem** and **anxiety about body fat**. Sufferers have a **poor self-image**. Even though they lose loads of weight, they still see themselves as fat and **continue starving themselves**.

2) Symptoms of anorexia are muscle wasting, depression, delayed sexual development, retarded growth and decreases in BMR and heart function. Severe anorexia can lead to **death by starvation**.

Practice Questions

Q1 What causes marasmus?

Q2 Name four symptoms of anorexia nervosa.

Q3 Name a deficiency disease caused by lack of vitamin D.

Q4 What percentage over recommended body weight is classified as obese?

Q5 Explain what PEM is.

Exam Questions

Q1 Describe the dietary causes of a named disease found mainly in:

　　a) LEDCs; [2 marks]

　　b) MEDCs. [2 marks]

Q2 Explain why we need vitamin A in our diet. [4 marks]

Supercalorificdiet-totallyatrocious... *(*in a Mary Poppins stylee)*

Remember that people in both developed and developing countries suffer from dietary illnesses through malnutrition — but the diseases are different. In LEDCs, they're usually deficiency diseases or starvation, and in MEDCs they're more to do with over-eating and having high-fat diets.

Gaseous Exchange

Two nice, straightforward pages on gaseous exchange and physical fitness. Deep breath... and away you go.

Lungs are Specialised Organs for Breathing

Mammals like humans exchange oxygen and carbon dioxide through their **lungs**. Lungs have special **features** that make them well-adapted to **breathing**:

Cartilage — rings of strong but bendy cartilage keep the **trachea** open.

Goblet cells — produce **mucus** to trap inhaled dust and other particles.

Cilia — **hairs** on the cells that line the trachea, bronchi and bronchioles. They **move** to push the mucus with trapped particles **upwards**, away from the lungs.

Smooth muscle — round the bronchi and bronchioles. **Involuntary** muscle contractions narrow the airways.

Elastic fibres — between the alveoli. Stretch the lungs when we breathe in and recoil when we breathe out to help push air out.

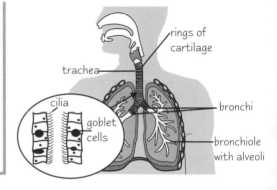

In Humans Gaseous Exchange Happens in the Alveoli

Lungs contain millions of microscopic air sacs called **alveoli**. They're responsible for gas exchange. They're so tiny, but so important — size ain't everything.

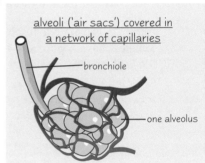

alveoli ('air sacs') covered in a network of capillaries

1) The huge number of alveoli means a **big surface area** for exchanging oxygen and carbon dioxide.

2) O_2 diffuses **out of** alveoli, across the **alveolar epithelium** (the single layer of cells lining the alveoli) and the **capillary endothelium** (the single layer of cells of the capillary wall), and into **haemoglobin** in the **blood**.

3) CO_2 diffuses **into** the alveoli from the blood, and is breathed out through the lungs, up the trachea, and out of the mouth and nose.

4) The alveoli secrete liquid called **surfactant**. This stops the alveoli collapsing by lowering the surface tension of the water layer lining the alveoli.

Alveoli have **adaptations** that make them a really good surface for gas exchange. They have the following features, which all **speed up** the **rate of diffusion**:

- **thin exchange surfaces** (alveolar epithelium cells are very thin);
- **short diffusion pathways** (the alveolar epithelium layer is only one cell thick);
- **a large surface area to volume ratio**;
- **a steep concentration gradient** between the alveoli and the capillaries surrounding them (see page 22).

TV is the Volume of Air in a Normal (Resting) Breath

dm^3 is short for decimetres cubed

There are two main terms that you need to know about breathing (or ventilation):

1) **Tidal Volume (TV)** is the volume of air in each breath — usually about **0.4 dm^3** per breath.

2) The **Vital Capacity** of the lungs is the maximum volume of air that can be breathed in or out.

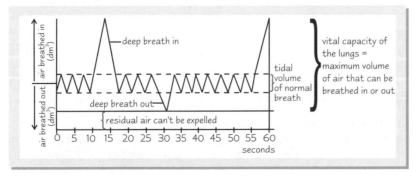

The <u>medulla oblongata</u> in the brain controls rate of breathing. Breathing rate changes depending on how much physical activity you're doing. When exercising you use more energy. Your body needs to do more <u>aerobic respiration</u> to release this energy (see p.50), so it needs more oxygen.

Fitness

Your **Pulse Rate** is a Measure of Your **Physical Fitness**

Pulse Rate / Heart Rate — Pulse rate is a measure of the heart rate. Blood is forced through the arteries by the heart contracting. You can feel your pulse where arteries are near the skin's surface, like in your wrist. During exercise the pulse (heart rate) speeds up. Afterwards it returns to resting rate.

Fitness is measured by how quickly the pulse returns to **resting rate**.
The quicker it returns, and the lower the resting pulse, the fitter you are.

There are **Three Terms** You Should Know About **Blood Pressure**

The performance of the heart can be monitored by recording changes in the pulse rate and in the blood pressure. So, you need to learn these three terms. The fun never ends...

1) **Systolic Pressure** — the pressure in the ventricles (heart chambers — see p.68) when heart muscle **contracts**.

2) **Diastolic Pressure** — the pressure when the ventricular muscle **relaxes**.

3) **Hypertension** — is when **blood pressure** is too **high**. This puts extra strain on the heart muscle and on the blood vessels, which causes heart problems.

Systolic blood pressure increases during exercise, so that oxygenated blood can be transported to the muscles more quickly. Diastolic blood pressure stays about the same.

Practice Questions

Q1 Draw a flow diagram of gas exchange in the alveoli.

Q2 Why is surfactant important?

Q3 Where is the rate of breathing controlled?

Q4 Name three ways the lungs are adapted for gas exchange.

Q5 How does a person's pulse rate relate to their physical fitness?

Q6 What is diastolic pressure?

Exam Questions

Q1 Explain why hypertension is a danger to health. [2 marks]

Q2 Explain why a specialised gas exchange system is required in humans. [4 marks]

You need to learn about TV — watching Eastenders doesn't count...

I may have been a bit optimistic when I said this was straightforward. OK, I was lying. There are loads of terms to learn, and it's tempting just to read them and hope for the best. But the safest way is to close the book and try to write them out, along with their meanings. Keep doing this until the cows come home and the fat lady sings — or, until you know them all.

Energy Sources and Exercise

Aerobic exercise doesn't necessarily involve large housewives in colourful spandex. It means any exercise that makes the cardiovascular and gas exchange systems of the body work harder.

There Are **Two** Types of **Respiration** — Aerobic and Anaerobic

When you exercise, your muscles need to be provided with energy so that they can keep working. This energy comes from respiration. There are two different types of respiration that happen in the body — aerobic and anaerobic. Anaerobic respiration releases a different amount of energy from aerobic respiration, and gives different end products.

(1) Aerobic Respiration —

1) Uses oxygen and produces waste CO_2 that's excreted through the lungs.

2) Releases more energy from each glucose molecule than anaerobic respiration.

> **glucose + oxygen $\longrightarrow$ carbon dioxide + water + ATP (contains energy)**

(2) Anaerobic Respiration —

1) It's less efficient at releasing energy.

2) It doesn't need oxygen to release energy.

3) It produces **lactic acid**, which builds up in the blood. This lowers the pH in muscle cells that are respiring anaerobically, which causes the pain known as **muscle fatigue**.

> **glucose $\longrightarrow$ lactic acid + ATP (contains energy)**

During Exercise an **Oxygen Debt** May Build Up

This is an immediate effect of exercise on the body.

During **vigorous exercise** the body demands more oxygen than is available:

1) In skeletal muscles, **aerobic** respiration changes to **anaerobic** respiration.

2) **Lactic acid** is produced, which builds up in the blood. Lactic acid is **toxic** and the body can only deal with **small amounts**. Too much lactic acid in the blood means exercise has to stop.

3) **Oxygen** is needed to get rid of the lactic acid. When exercise stops, more oxygen can be used for this job.

4) The oxygen needed to convert lactic acid to the chemical **pyruvate** is called the **oxygen debt**. This is why you keep **panting** after hard exercise — you're still **repaying** the oxygen debt.

Aerobic exercise is exercise using oxygen. If you pace yourself well you can have a good work-out without going into anaerobic respiration. Athletes train regularly to train their muscles to work harder whilst staying in aerobic respiration (which provides more energy). Aerobic exercise improves **ventilation** and makes the **circulatory system** more efficient.

The **Effects** of Exercise can be **Investigated**

Long term research has been conducted into the benefits of exercise — one way of doing this is by comparing people with different fitness levels:

1) Expose people with **different fitness levels** (e.g. an athlete compared to a person who does little exercise) to a measured exercise regime.

This type of experiment has shown that people who exercise regularly have improved heart and lung function.

2) Make sure that the fitness level is the only variable and keep everything else **constant,** — like age, sex, duration of activity, etc.

3) **Take measurements** such as:

- **tidal volume** and **vital capacity** of lungs;
- heart rate and blood pressure;
- body temperature;
- blood concentration of CO_2 and lactic acid.

4) Do this before, during and after exercise, using appropriate equipment — like a **spirometer** to measure the volume of air moving in and out of lungs.

5) **Analyse** results to work out things like **recovery rates**, total oxygen consumed during exercise, etc.

Energy Sources and Exercise

You Need to do **Exercise** to Stay Healthy

Recent studies have found that **two thirds** of adults in the UK don't do enough exercise to stay healthy. Frequent, moderate exercise keeps the body working at its maximum efficiency, and increases the amount of the chemicals in the brain which make you feel good.

Look how happy you could be.

Exercise makes you happy:

The **pituitary gland** releases an increased level of ß-endorphin into the blood during exercise. Endorphins belong to a family of chemicals that includes **morphine** — used to relieve pain. It's thought that ß-endorphin is responsible for the "jogger's high" — a feeling of well-being after exercise.

There are **Three Types** of Exercise

Each type has its own particular long-term benefits:

1) **Aerobic Exercise** — activities that increase **heart rate** and **lung efficiency**, e.g. walking, running, swimming or cycling.

Benefits
- Keeps heart muscle healthy.
- Increases **oxygen delivery** to body cells.
- Improves lung efficiency.
- Decreases **cholesterol**.
- Improves resistance to disease.
- Reduces excess weight.

To see a significant, sustained improvement in your aerobic fitness, it's recommended you exercise three times a week for at least twenty minutes. This should be at 70% of your maximum capacity.

2) **Muscle Strengthening Exercise** — works the muscles against resistance.

Benefits
- Improves muscle control of skeleton and body shape.
- Improves strength.

3) **Flexibility Exercises** — stretching exercises like **yoga** and **pilates**.

Benefits
- Improves posture, muscle tone and body shape.
- Also said to improve mental fitness such as the ability to concentrate or relax.

Practice Questions

Q1 What is the main respiratory substrate used by animals?

Q2 Write out the word equations for aerobic and anaerobic respiration.

Q3 What causes muscle fatigue?

Q4 Would you expect a trained athlete to consume more or less energy than other people during exercise? Why?

Q5 Name three types of exercise and give the benefits of each.

Exam Questions

Q1 Explain what happens during vigorous exercise in terms of respiration. [4 marks]

Q2 Explain why regular exercise can prolong an active life. [5 marks]

First revision, now you're nagging about exercise — leave me alone...

Actually, exercise can help you revise — put down your Revision Guide for a bit (the first and last time I'll ever say that) and do some gentle exercise. When you come back to it, you'll find it easier to concentrate and you'll remember things more quickly. And now you know what causes those vicious pains if you exercise a bit too much. Marvellous.

Effects of Smoking

Don't worry, I'm not about to launch into a lecture. Everyone has a choice to make about whether or not they're going to start smoking. All these pages do is tell you the scientific stuff about it that you need to know to pass your exam.

Smoking Damages the Body

Some of the health problems caused by smoking include:

1)   Toxic chemicals in the tar can cause cancer

Scientific evidence was first used to show the link between smoking and **lung cancer** in the **1950s**. **Tar** from cigarette smoke collects in the lungs. It's full of toxic chemicals, some of which are **carcinogens** (cause cancer). Carcinogens make mutations in the DNA of the **alveoli** more likely. If this happens, cell division goes out of control and **malignant tumours** form. Lung cancer is one of the main causes of early deaths in the UK.

2)   Inhaling smoke can cause chronic bronchitis

The smoke destroys the **cilia** on the epithelial tissue lining the trachea. It also irritates the **bronchi and bronchioles**, which encourages mucus to be produced. Excess mucus can't be cleared properly because the cilia are damaged. It sticks to air passages causing smoker's cough and **chronic bronchitis**. Micro-organisms multiply in the extra mucus. (Not a happy thought if you're snogging a smoker.)

3)   Emphysema is a breathing disorder caused by reduced gas exchange

People suffering from emphysema have a reduced surface area for gas exchange, and a loss of **elasticity** in their lungs. A lot of smokers develop it eventually, because chemicals in tobacco smoke destroy **fibrous elastin tissue** in alveoli. This means the alveoli can't expand as air enters the lungs. The smoke also increases the number of protein-eating **phagocyte cells** (see p.58), which digest the alveoli tissue. This reduces the surface area for gas exchange. Sufferers gasp for breath because their lungs can't absorb enough oxygen, and they might need an oxygen cylinder to help them live longer.

Smoking can cause Artery and Heart Disease

Smoking is a risk factor in **atherosclerosis** (a disease caused by narrowing of the arteries), coronary heart disease and strokes.

This is because tobacco smoke —

1)   Destroys **elastin**, so blood vessel elasticity is reduced. This leads to narrowing of the arteries.

2)   Encourages **blood clots** to form more easily.

3)   Contains nicotine, which activates **adrenalin**. This raises blood pressure and puts more strain on heart muscle. (Nicotine is also the substance that smokers become addicted to.)

4)   Contains **carbon monoxide**, which combines irreversibly with haemoglobin — so there's less haemoglobin available for oxygen transport.

A **stroke** can be caused when an artery narrows or become blocked by a clot. This reduces oxygen supply to the brain and causes loss of certain functions, like memory, and can be fatal.

Coronary heart disease is caused by increased heart rate and slower delivery of oxygen to the heart due to blocked arteries. This means the heart can't keep functioning properly.

Thromboses (blood clots) happen in narrowed arteries. They can block the arteries, which means oxygen fails to reach heart muscles.

Effects of Smoking

Statistics show Smokers are more likely to Die Early

Epidemiological (see p.42) and experimental evidence link cigarette smoking with disease and early death:

1) **Richard Doll's** statistical study in the 1950s first showed the link between smoking and lung cancer.

2) **Animal experiments** conducted in the 1960s showed that tumours developed when tobacco smoke was present in the lungs.

3) **Chemical analysis** of the chemicals in tobacco smoke has proved the presence of **carcinogenic substances**.

4) **Epidemiological** studies have shown that **passive smoking** (breathing in other people's smoke) increases the chances of developing lung cancer. They have also shown that the risk is reduced if a smoker stops smoking.

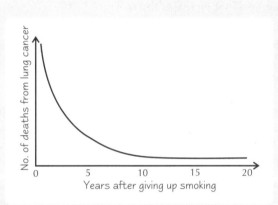

The **graph** shows the results from epidemiological studies on the mortality of ex-smokers. It shows that smokers are less likely to die from lung cancer once they've given up smoking.

Jeff — living proof that smoking isn't necessarily cool.

Practice Questions

Q1 What is a carcinogenic substance?

Q2 List three effects of tobacco smoke on the body.

Q3 List three diseases that smokers are more likely to suffer from than non-smokers.

Q4 Why does smoking increase the likelihood of blood clots?

Q5 Give three types of study that have shown that smoking causes cancer.

Exam Questions

Q1 Give four health reasons why it is sensible to ban smoking in restaurants, theatres and other public places.   [4 marks]

Q2 'Emphysema is damage to the exchange surface of the lungs'. Explain how smoking causes this damage.   [4 marks]

Disease, death, mucus — such a delightful subject...

It might not be pleasant, but I'm afraid it's all true and you're going to have to learn it. This is another thing that often comes up. Maybe the examiners are trying to catch out the people who quickly turned the page so they weren't reminded about what they're doing to their lungs. Don't be shy, smokers — these pages are all about you and your body. Fame at last.

Prevention and Cure

Most diseases can either be prevented, or cured.
Prevention is better than cure as my old mum used to say — but what did she mean...

CHD is an example of a Disease with Prevention and Cure

Coronary Heart Disease (**CHD**) is when not enough oxygen gets to the heart so it can't function efficiently. If the blood supply to an area of the heart is cut off then the heart muscle can't respire and fails. This causes a **heart attack**.

There are many factors which affect your chances of getting CHD:

1) Some people are **naturally more prone** to heart disease, for example if they've inherited high blood pressure or high cholesterol levels in their blood.

2) The **width of arteries** carrying blood to the heart affects how much oxygen is provided to the heart muscle. **Cholesterol** clogs up arteries, making them **narrower**, so less blood can get to the heart.

3) **Smoking** is another **risk factor** that increases the chance of suffering from coronary heart disease.

4) People with **diabetes mellitus** or high **blood pressure** are also more likely to develop heart disease.

5) Excess alcohol intake increases blood pressure, so that increases the risk too.

6) One important inherited factor is a person's **sex** —
men are three times as likely as women to suffer from coronary heart disease.

Some risk factors are preventable:

1) Over-nutrition and high-fat diets lead to **excess weight**. This should be avoided, because otherwise the heart muscle is **put under strain** pumping blood to the extra body cells.

2) **Diets** with too much cholesterol should be avoided to prevent narrowing of arteries.

3) **Regular exercise** is also needed to keep the heart muscle strong.

4) Cutting out the '**self-inflicted**' risk factors like smoking and alcohol abuse. There are many ways to help people give up smoking, from nicotine replacement patches to hypnosis.

Death Rate from Coronary Heart Disease Varies Around the World

Death rates aren't the same in every country. For example, the number of deaths from coronary heart disease is higher in Finland and Britain than in France or China. This difference in the death rate is mainly due to the factors below:

1) Differences in the **inheritance** of disease — inherited factors aren't preventable, but an awareness of family health history does help people make better decisions about their lifestyle.

2) **Diet** varies between countries. Some countries (like Britain and Finland) have, overall, more fatty diets than other countries.

3) **Attitudes** towards drinking and smoking vary between countries.

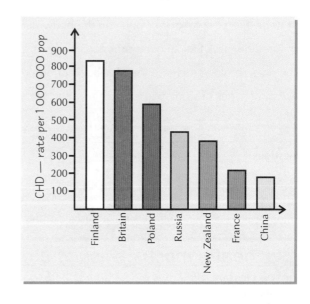

Prevention and Cure

By-pass Operations Improve Heart Function

A **coronary artery bypass** operation only relieves the **symptoms** of coronary heart disease
— the causes can't be treated by surgical methods. By-pass surgery is used when
coronary arteries can't be unblocked. Sections of **veins** taken from a leg are attached to
the heart. These are used to **link blood vessels**, by-passing the blockage.

Blood flow is increased, improving the supply of oxygen and nutrients to **heart muscle**.

Transplanted Hearts can be Rejected

If a person's heart is severely damaged, a **transplant** can be their only chance of survival. Hearts become
available for transplant usually after **accidents** such as car crashes, which kill someone yet leave their heart
undamaged. But hearts for transplants aren't plentiful, and even if one were available, it might be **rejected**:

1) The function of **white blood cells** in the immune system is to destroy
   'foreign' tissue like bacteria and viruses.

2) Unfortunately, heart tissue from another person is identified as foreign by the immune system.

3) Heart transplant patients are treated with drugs to **prevent rejection** of new heart tissue.

4) If these fail the heart tissue is **rejected** — it's attacked by the immune system.

5) If another transplant isn't available at once, the patient will die.

Have a look at page 68
for more on the heart.

Practice Questions

Q1   List four risk factors which lead to coronary heart disease.

Q2   Are any of the risk factors you have listed preventable factors? What does this mean?

Q3   Why is there a danger of rejection after a heart transplant?

Q4   Why might the death rate from heart disease be lower in China than Britain?

Exam Question

Q1   Describe what happens in a coronary artery by-pass operation and explain how this can
      help a patient suffering from coronary heart disease. Why are such patients not simply
      given a new heart in a transplant operation?   [8 marks]

Take your pick — prevention or cure...

*This isn't a hard topic, but don't just rush through it or you'll forget little details. Like why heart transplants aren't always
straightforward, and which factors affect your chances of having CHD. So, same old, same old — keep going over the
page and scribbling down all the details until you know them all, without having to look.*

Infectious Diseases

Another joyful page. Lots more lovely bacteria, viruses and disease for you to get your teeth into. Yummy.

AIDS *stands for* Acquired Immune Deficiency Syndrome

AIDS is caused by the **Human Immunodeficiency Virus** (HIV). So far over **30 million** people have been infected with HIV, and there's no vaccine available to stop it spreading more. There's no known cure for AIDS either.

1) HIV is a **retrovirus** — it injects its own RNA into the **nuclei** of host cells.
2) HIV can remain **dormant** for many years, then it replicates and kills the host cells.
3) The host cells targeted by the virus are the cells responsible for fighting pathogens (**T-cells**), so the infected person ends up with a **weakened immune system**.
4) This weakened immunity is called **AIDS**. It means the body can't defend itself against diseases like **pneumonia**, which it would usually recover from.

HIV is transmitted in body fluids

Body fluids include semen (so you could be infected during unprotected sex), and **blood** — so drug users who share needles are at risk. Babies can be born with the disease after infection through the **placenta** if their mother has HIV. It can also be transmitted through **breast milk**.

Attempts are being made to stop the spread of HIV

Donated blood is **screened** for HIV to reduce transmission through **blood transfusions**. Using condoms during sex helps to stop you catching HIV (and other unpleasant diseases). Free **needle exchange centres** have been set up to provide drug addicts with clean needles, and there are drugs available to reduce the risk of a pregnant woman infecting her unborn child.

The main preventative measure for HIV and AIDS is education — leaflets were delivered throughout the UK in the 1990s with the message "Don't Die of Ignorance."

Malaria *is Caused by* Plasmodium

Plasmodium is a **protozoan** — a eukaryotic, single-celled organism (see p.6). It's carried by **mosquitoes**, which feed on blood from animals like humans. The mosquitoes are **vectors**, meaning they carry the disease without getting it themselves. They transfer the malarial parasite (*Plasmodium*) from one animal to another by inserting it into animals' blood vessels when they feed on them. Mosquitoes infect up to **500 million** people a year. Malaria kills **3000 children a day**, and a total of **1 million people** a year.

The main prevention measures are —
1) **Drain** the areas of water where mosquitoes lay their eggs.
2) **Spray** these areas with **insecticides**.
3) **Introduce fish** into the water to eat mosquito larvae.
4) **Protect people** from mosquitoes by using **mosquito nets** and insecticides.

Biological control can be used to reduce infection rates

Scientists sterilise male mosquitoes using radioactive sources
↓
Sterilised males released into the malarial area
↓
Females mate with sterilised males
↓
No reproduction occurs
↓
Fewer mosquitoes produced
↓
Malaria disease is reduced

Tuberculosis *(TB) Kills Over* 2 Million People *a Year*

Tuberculosis is caused by the bacterium *Mycobacterium tuberculosis*, and spreads rapidly by **droplet infection** (where bacteria are inhaled from the air).

The bacteria can live for years in the lungs before they become active. This **dormant** period prevents **contact tracing** (i.e. you don't know **when** you caught the disease so you don't know **who** you caught it from or who might have caught it from you) — so it's very difficult to eradicate the disease completely.

TB can be prevented with the **BCG vaccine** and controlled using the antibiotic **Streptomycin**, but **resurgence** often happens because of **resistance** to the drug.

The number of reported cases of TB has been rising steadily in the UK — from **5010** in 1990, to **6379** in 2000. A major group affected is **homeless people**. Improved housing conditions with less overcrowding could help.

Infectious Diseases

Cholera is Caused by Vibrio cholerae, a Water-borne Bacterium

Cholera is an infectious disease which causes severe diarrhoea. It is initially transmitted by **drinking contaminated water**. An infected person can then pass it on by handling food with unclean hands.

Some areas of the world regularly suffer from **epidemics** due to poor housing and sanitation. Investment in sewage treatment works and **chlorination** of drinking water would help eradicate the disease.

Poorer Countries Need Help to Control these Diseases

LEDCs (see p.43) often lack the resources needed to treat and prevent the spread of these diseases. Money is needed for things like:

1) improved water treatment (to prevent **cholera**);
2) elimination of mosquito breeding areas (to prevent **malaria**);
3) improved housing conditions, vaccinations and antibiotics (to prevent **TB**);
4) education (to prevent **AIDS**).

The **World Health Organisation** helps eliminate diseases around the world. One of the biggest problems is the spread of **HIV in Africa**. Africa has the highest incidence of HIV per head of its population in the world.

Antibiotics are used to Treat Infectious Bacterial Diseases

Antibiotics are chemicals produced by some **micro-organisms**. They can be used to inhibit or destroy **harmful bacteria**, because they interfere with the bacteria's metabolism without interfering with ours.

1) The first antibiotic to be discovered was penicillin. Most antibiotics were originally produced from the *actinomycete* fungus that occurs naturally in soil.

2) Antibiotics can be used to treat most bacterial diseases in humans, from **typhoid** to sore throats. But some bacteria have developed a **resistance** to antibiotics due to **misuse** or **overuse** of the drugs. For example, if a person doesn't finish their full course of antibiotics, the most resistant bacteria will be left to reproduce. Bacteria reproduce **asexually**, so their offspring will be clones and will therefore also be resistant to the antibiotic. This may result in a resistant strain. Uh-oh.

3) Antibiotics are made in large quantities in **fermentation reactions**. The biochemical industry constantly researches and develops new strains.

Practice Questions

Q1 How is TB spread?

Q2 Which of these diseases is caused by water contamination: Malaria, Cholera, TB?

Q3 How does increasing fish stocks in lakes reduce malaria?

Q4 List four ways money could be spent to eradicate TB, cholera and AIDS in LEDCs.

Q5 Blood transfusions are used regularly in hospitals. How are patients protected from HIV?

Exam Questions

Q1 Explain why contact tracing is important in treatment of infectious disease. [3 marks]

Q2 Give three ways that HIV is transmitted. [4 marks]

TB or not TB? That is the question

...and 'hopefully not,' is the answer. I don't know — first cancer and emphysema, then heart disease, and now TB and AIDS. Don't they realise AS levels are depressing enough without all this? Don't bother learning these pages because we're all going to die anyway. Dooooom. Glooooom.

The Immune System

The immune system protects the body from pathogens (organisms that cause disease). It helps the body recognise them as foreign, and destroys them with a mixture of funny-sounding cells, like phagocytes and lymphocytes. Hurrah.

Antigens and Antibodies Know What Should and Shouldn't be in Your Body

Antigens and **antibodies** work together to let your cells distinguish 'self' from 'non-self' — they know when something is in your body that shouldn't be.

1) **Antigens** are large, organic molecules found on the surface of **cell membranes**. Every individual has their own **unique** cell surface molecules, which the immune system recognises. So when a **pathogen** like a bacterium invades the body, the antigens on its cell surface are identified as **foreign** by the immune system.

2) When the body detects foreign antigens, it makes **antibodies** — **protein molecules** that bind to specific antigens, giving an **antigen-antibody complex**.

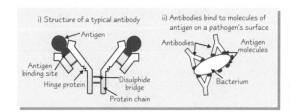

3) Antibodies have **two binding sites**, so they can attach to two antigen molecules. They deal with pathogens by **clumping** or **linking** them together to make it easier for them to be engulfed by **phagocytes**. They can also **rupture** foreign cells, (which kills them) and inactivate any **toxins** they produce.

Phagocytes Get Rid of Cell Debris and Bacteria

1) **Phagocytes** are large white blood cells that encircle and **engulf** pathogens — a process called **phagocytosis**. Phagocytes use phagocytosis to clear away the antigen-antibody complexes formed in an **immune response**.

2) **Macrophages** and **neutrophils** are types of phagocytes. They're both made in the **bone marrow**. Neutrophils are found in the **blood** and macrophages in **tissues**.

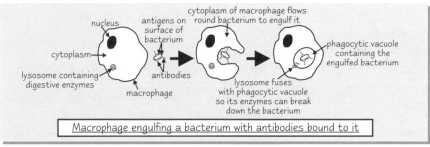

Macrophage engulfing a bacterium with antibodies bound to it

Lymphocytes Control our Immune Response

Lymphocytes are white blood cells with **glycoproteins** on their surface to **recognise foreign antigens**. The two main types are: **B-lymphocytes** (**B-cells**) that mature in the **b**one marrow; and **T-lymphocytes** (**T-cells**) that mature in the **t**hymus gland.

Our bodies carry out **two types** of immune response:

1) **The CELLULAR RESPONSE** — This uses **T-cells**, either to attach to foreign antigens and attack pathogens directly, or to trigger the **humoral response** (see below). There are **four types of T-cell**.

- **Killer or cytotoxic T-cells** — stick to a pathogen's antigens and secrete substances that kill the pathogen.
- **Helper T-cells** — help the **plasma cells** produced by **B-lymphocytes** to secrete antibodies, and secrete substances that attract **macrophages**.
- **Memory T-cells** — Keep the ability to recognise a particular antigen in the future. Allows the immune system to respond very quickly to a future attack by **the same pathogen**.
- **Suppressor T-cells** — suppress activity of killer T-cells and B-cells when antigens have been destroyed.

2) **The HUMORAL RESPONSE** — This uses **B-cells** to release antibodies. Each **B-cell** has a **specific antibody** on its membrane. When they meet a complementary foreign **antigen**, the antigen binds to the antibody's **receptor site**. This stimulates the B-cell to release the antibody into the blood. The B-cell then divides by mitosis, producing many clones, called **plasma cells**. These cells release large quantities of the same antibody, hopefully killing the pathogen. Some of the clones become **memory B-cells**, which provide us with immunity against a pathogen if it enters the body again. The next page has more on memory cells.

The Immune System

The **Immune Response** for Antigens can be **Memorised**

The first time a particular antigen enters the body the **immune response** is slow, because there aren't many B-cells that can make the antibody needed to bind to it. This is the **primary response**. Eventually the body will probably produce enough of the right antibody to overcome the infection, but while it's doing this the infected person will show **symptoms** of the disease. After being exposed to an antigen, both T- and B-cells produce **memory cells**. Memory T cells record the **invading antigen**, and memory B- cells record the '**recipe**' **for antibodies** to fight them.

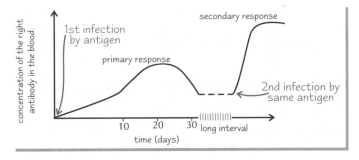

If the same pathogen enters the body again, these memory cells can produce the right cells and antibodies for fighting the pathogen very quickly.
This is known as the **secondary response**. It often gets rid of the pathogen before you begin to show any symptoms.

The **Two Immune Responses** can be Summarised as Shown

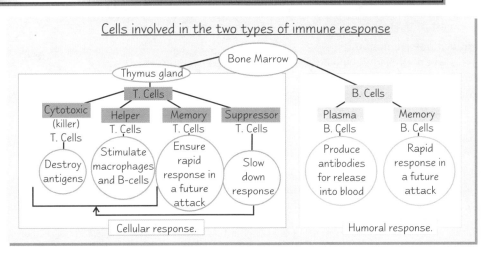

Practice Questions

Q1 What are antigens?

Q2 What is an antibody?

Q3 Where do B-lymphocytes come from, and what is their function?

Q4 Name the four types of T-lymphocyte.

Q5 What is phagocytosis?

Exam Questions

Q1 How does the cellular response protect the body from disease? [8 marks]

Q2 Explain how antibodies are produced, and describe their function. [6 marks]

The humoral response? I'm sorry, but I don't get the joke...

I'm not amused by the names of some of these cell types either. I'm sure we're all very grateful to these macrophages, neutrophils and cytotoxic T-lymphocytes for doing their bit to defend us from the attacking hordes. But why couldn't they have names like Bill, Jim and Dave?

Immunity and Vaccinations

This page is pretty darn interesting. It's nice to know there's a reason for hay fever making your life hell all summer, and to understand why people sticking needles in you is a good thing.

Immunity can be Active or Passive

Active immunity — the immune system makes **antibodies** of its own after being stimulated by an **antigen**. There are two different types of active immunity:

1) **Natural active immunity.**
   This the type of immunity you have after you catch a disease and it promotes an **immune response** (see pages 58 and 59).

2) **Artificial active immunity.**
   This is the type of immunity you have after you're given a **vaccination** containing a harmless dose of **antigen** (see the next section).
   Your body makes **antibodies** against it, but you don't get the disease.

Passive immunity — uses antibodies made by **another** organism. The immune system doesn't develop any antibodies of its own. Again, there are two types:

1) **Natural passive immunity.**
   An example of this is the type of immunity that a baby has from its mother.
   Antibodies are passed from mum to baby through the placenta and in the breast milk.

2) **Artificial passive immunity.**
   This is when a person is injected with **antibodies** from someone else, rather than with a harmless dose of antigen. An example is a **tetanus jab**, which contains antibodies against tetanus toxin (collected from blood donations).

The main advantage of active immunity is that it involves producing **memory cells,** which allow a rapid **secondary response** to the same pathogen if it's encountered again.

Memory cells aren't produced in passive immunity, so it's only **short-term** protection.
But it's **immediate** — with active immunity the primary immune response takes a while.
With natural active immunity, this means you develop symptoms of the disease, and it could even be fatal.

Vaccines are Made From Harmless Antigens

Vaccines contain harmless **antigens** that are injected or swallowed.
The dose is very small, and a killed or weakened form of the pathogen is used.
The body then starts making **antibodies** against the antigens in the vaccine.
T-lymphocyte and B-lymphocyte **memory cells** store the information about the antigen and its required antibody. If the real pathogen then enters the body, the immune system can work fast enough so that you don't get ill.

The **World Health Organisation** (WHO) uses vaccination as a major weapon in the fight against disease.
A vaccination programme, coordinated by the WHO, **eradicated smallpox** from the world. Yay!

This was possible because —

1) Only one vaccine and no **boosters** (extra doses of the vaccine needed after the first one) were needed, so it was cheap to make.

2) There was only **one strain** of the pathogen, and it couldn't infect animals as well as people.

3) The disease was easily **diagnosed**, and everyone who got infected showed the symptoms.

4) So when the disease started dying out, there was no pathogen still hidden in the population or in animals to infect people again in the future.

Immunity and Vaccinations

Measles, TB, Malaria and Cholera Can Be Vaccinated Against Too...

... but haven't been eradicated yet.

Some of the reasons are —

1) The pathogens exist in many strains, which keep changing by **mutation**, e.g. the influenza virus.

2) Some live in animals too, e.g. **malaria** lives in mosquitoes.

3) Some invade the **gut** where the immune system doesn't work as well, like **cholera**.

4) It can be hard to organise vaccination programmes — e.g. for **measles** in Less Economically Developed Countries, vaccinations aren't always readily available, so babies often can't be vaccinated soon enough after birth to give them protection when they're most vulnerable.

An Allergy is a Reaction to a 'Harmless' Substance

Allergies are over-reactions by the immune system to harmless antigens, called **allergens**. People with **hay fever** are sensitive to pollen. When they breathe it in the body over-reacts and increases production of chemicals like **histamine**. This causes **inflammation** of the affected area, excessive mucus secretion and constriction of the **bronchi** (so you're swollen, snotty and can't breathe). **Anti-histamine** is used to treat hay fever, insect bites and skin allergies.

Asthma can be caused by an allergic reaction too. The allergen is the **house dust mite** and its faeces. Asthma causes inflammation in the **bronchioles**. This obstructs the airways, making it hard to breathe.

Maurice just couldn't understand why his hay fever was playing up in the middle of January.

Practice Questions

Q1 What is the difference between passive and active immunity?

Q2 Which type of immunity is passed from mother to unborn baby — passive or active?

Q3 Give three features of smallpox that allowed it to be eradicated using a vaccination programme.

Q4 How is asthma caused?

Q5 What does anti-histamine do?

Exam Question

Q1 Smallpox is a disease that was eradicated using a vaccination programme.
Many other diseases can also be vaccinated against, but have not yet been eradicated.
Discuss the possible reasons why such diseases are less susceptible to vaccination programmes.   [7 marks]

So asthma's an allergy to mite poo — you learn something every day...

Or in your case you learn 492 new things. So perhaps you're not as excited about mite poo as I am. Make sure you learn all the facts on these pages. Spot check — how is passive immunity different from active immunity? What's an allergy? If you can't answer these questions then back you go to the start of page 60. Wachoo.*

*The sound of a whip cracking, in case you were wondering.

Transport Systems

These pages are about transport. Big organisms need exchange surfaces to get enough oxygen, food and the rest of it inside them. Once it's in there, they have to be able to move it about to where it's needed — so they use transport systems.

Smaller Animals have Higher Surface Area : Volume Ratios

A mouse has a bigger surface area **relative to its volume** than a hippo. This can be hard to imagine, but you can prove it mathematically. Imagine these animals as cubes:

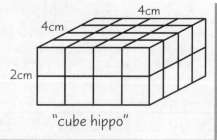

The hippo could be represented by a block with an area of 2 cm × 4 cm × 4 cm.

Its **volume** is 2 × 4 × 4 = **32 cm³**

Its **surface area** is 2 × 4 × 4 = 32 cm² (top and bottom surfaces of cube)

 + 4 × 2 × 4 = 32 cm² (four sides of the cube)

Total surface area = **64 cm²**

So the hippo has a **surface area : volume ratio** of 64 : 32 or **2 : 1**.

"cube hippo"

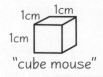

"cube mouse"

Compare this to a mouse cube measuring 1 cm × 1 cm × 1 cm

Its **volume** is 1 x 1 x 1 = **1 cm³**

Its **surface area** is 1 x 1 x 6 = **6 cm²**

So the mouse has a **surface area : volume ratio** of 6 : 1

← The cube mouse's surface area is six times its volume. The cube hippo's surface area is only twice its volume. Smaller animals have a bigger surface area compared to their volume.

Multicellular Organisms need Transport Systems

All cells need energy — most cells get energy via **aerobic respiration**. The raw materials for this are **glucose** and **oxygen**, so the body has to make sure it can deliver enough of these to all its cells. In single-celled creatures, these materials can **diffuse directly** into the cell across the cell surface membrane. The diffusion rate is quick because of the small distances the substances have to travel (see p.22).

In **multicellular** animals, diffusion across the outer membrane is too slow for their needs. This is because:

1) some cells are **deep within the body**;

2) they have a **low surface area to volume ratio**;

3) they have a **high metabolic rate**, which means they respire quickly, so they need a **constant supply** of glucose and oxygen.

4) they have a **tough outer surface**.

So multicellular animals need **transport systems** to carry raw materials from specialised **exchange organs** (like the lungs in humans, see p.48) to their body cells. In mammals this is the **circulatory system** (see p.69), which uses **blood** to carry glucose and oxygen around the body. It also carries **hormones**, **antibodies** (to fight disease) and **waste** like CO_2.

Surface area is also important for **body temperature**. Animals release heat energy when their **cells respire** and lose it to the **environment**. So **small animals** with high S.A. : volume ratios, like mice, have to use lots of energy just **keeping warm**, while **big animals** with low S.A. : volume ratios, like elephants, are more likely to **overheat**. Again, it's the job of the blood to help get rid of heat, by carrying warm blood close to the surface of the skin where it can lose heat to the environment. That's why you go red when you're too hot.

Animals with low S.A. : volume ratios have evolved adaptations to increase surface area. For example elephants have big, flat ears to increase their surface area to help heat escape from the body quickly.

Blood Vessels

Blood Vessels *Transport Substances Round the Body* in *Multicellular Organisms*

The three main types of blood vessel that you need to know about are **arteries**, **capillaries**, and **veins**.

1) **Arteries** are about 2mm in diameter. They carry blood under high pressure **from** the heart **to** the rest of the body. Arteries have **thick, strong** walls to cope with the **high pressure**. The walls are **elastic** so they can expand when the pumping of the heart causes surges of blood. All arteries carry **oxygenated** blood except the **pulmonary arteries**, which take deoxygenated blood to the lungs. Smaller arteries (**arterioles**) direct blood to different areas of demand by contracting and relaxing.

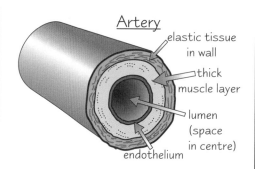

Artery
- elastic tissue in wall
- thick muscle layer
- lumen (space in centre)
- endothelium

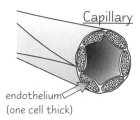

Capillary
- endothelium (one cell thick)

2) Arteries branch into **capillaries**, which are the **smallest** of the blood vessels (~8μm diameter). Substances are exchanged between cells and capillaries. Their walls are only **one cell thick** to allow efficient **diffusion** of substances (e.g. glucose and oxygen) to occur near cells. At their other end, capillaries merge into veins (see below). Networks of capillaries in tissue are called **capillary beds**.

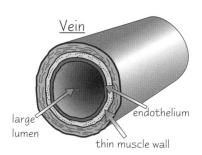

Vein
- large lumen
- endothelium
- thin muscle wall

3) **Veins** take blood under low pressure back **to the heart**. They're **wider** than equivalent arteries (diameter = 10 mm), with very little elastic or muscle tissue. Veins contain **semilunar valves** to stop the blood flowing backwards (see p.68). Blood flow through the veins is helped by contraction of the **body muscles** surrounding them. All veins carry **deoxygenated** blood (because oxygen has been used up by body cells), except for the **pulmonary veins**, which carry oxygenated blood to the heart from the lungs.

Practice Questions

Q1 Why don't single-celled organisms need transport systems?

Q2 What are the 3 main types of blood vessel?

Q3 Do arteries mainly carry oxygenated or deoxygenated blood?

Q4 What is the smallest type of blood vessel?

Exam Question

Q1 "Structures involved in transport are specialised for their function." Describe the relationship between the structure and function of the main three types of blood vessel found in the human body. [6 marks]

If blood can handle transport this efficiently, the trains have no excuse...

Four hours I was waiting at Preston this weekend. Four hours! Anyway, you may have noticed that biologists are obsessed with the relationship between structure and function, so whenever you're learning the structure of something, make sure you know how this relates to its function. Like the veins, arteries and capillaries on this page, for example.

Blood, Tissue Fluid and Lymph

For some people the word 'blood' is enough to make them cringe. Strange really, seeing as it's the substance that keeps us all alive. So sorry to all you blood-phobics out there, because these pages are all about the stuff.

Blood Contains Blood Cells, Platelets and Plasma

Blood is a **specialised tissue** that's composed of 45% **blood cells** and **platelets** suspended in a liquid called **plasma** (55%). Plasma is mainly **water**, with various nutrients and gases dissolved in it. The main role of blood is **transporting substances** around the body **dissolved** in the **plasma**. Substances move into and out of the plasma through **blood capillaries** at exchange surfaces.

Blood is used to transport lots of different substances around the body:

- **Hormones**
- **Oxygen**
- **Carbon dioxide**
- **Nutrients** from digestion —
  e.g. glucose, amino acids, fatty acids and mineral ions.
- **Urea**
- **Sodium ions**
- **Antibodies**

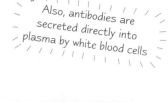

Also, antibodies are secreted directly into plasma by white blood cells

George had never really had a problem with the sight of blood.

Blood Cells are Adapted to Specific Jobs

1) **Red blood cells** are responsible for absorbing **oxygen** and transporting it round the body. They're made in the **bone marrow** and are very small.

   * They have **no organelles** (including no nucleus) to leave more room for **haemoglobin**, which carries the oxygen.
   * They have a large surface area due to their **bi-concave disc** shape. This allows O_2 to diffuse quickly into and out of the cell.
   * They have an **elastic membrane**, which allows them to change shape to squeeze through the small blood capillaries, then spring back into normal shape when they re-enter veins.

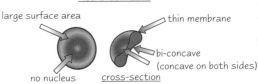

Red Blood Cells

large surface area / thin membrane / no nucleus / cross-section / bi-concave (concave on both sides)

The oxygen that red blood cells transport enters blood capillaries from the **alveoli** in the **lungs**. Make sure you're clear on gaseous exchange in the alveoli (look back to p.48 for more on this).

2) **White blood cells** are larger than red blood cells but there are fewer of them in blood. They are responsible for fighting disease (see p.58).

   There are two main types of white blood cell:

White Blood Cell	Diagram	Function	Structure
phagocytes		• engulf pathogens and micro-organisms and digest them (phagocytosis)	• elongated nucleus and flowing cytoplasm enables them to squeeze through gaps between cells to move to the site of infection in body tissues • contain many lysosomes to aid digestion
lymphocytes		• produce antibodies to prevent disease	• large nucleus

Blood, Tissue Fluid and Lymph

Tissue Fluid is Formed from Blood Plasma

Tissue fluid surrounds the cells in tissues — providing them with the conditions they need to function.
Tissue fluid is made from substances which leave the plasma from the blood capillaries.
The substances are squeezed out through the capillary walls due to the higher pressures inside.

Unlike blood, tissue fluid **doesn't** contain **red blood cells** or **big proteins**, because they are
too large to be pushed out through the capillary walls. It does contain smaller molecules,
e.g. oxygen, glucose and mineral ions. Tissue fluid helps cells to get the oxygen and glucose
they need, and to get rid of the CO_2 and waste they don't need. Waste substances either move
back into the blood, or drain into the lymph vessels ...

Lymph is Formed from Excess Tissue Fluid

Lymph is a fluid that forms when excess tissue fluid drains into the lymphatic vessels, which lie close to blood capillaries.
Lymph takes away waste products from the cells. It then travels through the **lymphatic system** and eventually enters blood plasma.

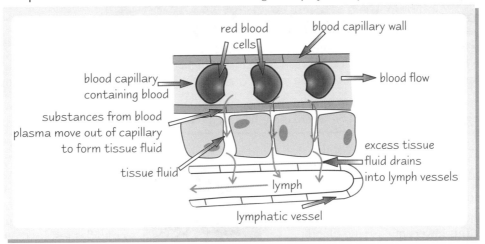

Lymph is similar to tissue fluid, except it contains less glucose and more fats, proteins and white blood cells,
which it picks up at **lymph nodes** as it travels through the lymphatic system.

Practice Questions

Q1   What is blood composed of?

Q2   Name 5 substances that are transported in the blood.

Q3   What do red blood cells do?

Q4   Describe three ways red blood cells are adapted for their function.

Q5   What are phagocytes and what is their function?

Q6   What are all cells surrounded by?

Q7   What makes lymph different from tissue fluid?

Exam Question

Q1   Explain how the structure of phagocytes helps them to fight disease in the body.   [3 marks]

That's the end of this bloody topic — and you can't get me for swearing...

*It's time to conquer any fears of blood and start appreciating it for the amazing tissue it is. Learn its functions, and which
bits do each function — plasma does transport, red blood cells do oxygen and white blood cells do disease-fighting.
If I were a blood cell I'd be a great phagocyte warrior, instilling fear into the nuclei of pathogens everywhere. Aaanyway.*

Haemoglobin and Oxygen Transport

Aaagh, complicated topic alert. Don't worry though, because your poor, over-worked brain cells will recover from the brain-strain of these pages thanks to oxyhaemoglobin. So the least you can do is learn how it works.

Oxygen *is Carried Round the Body as* Oxyhaemoglobin

Oxygen is carried round the body by **haemoglobin** (Hb), in red blood cells. When oxygen joins to it, it becomes **oxyhaemoglobin**. This is a **reversible reaction** — when oxygen leaves oxyhaemoglobin (**dissociates** from it), it turns back to haemoglobin.

$$Hb + 4O_2 \rightleftharpoons HbO_8$$
Haemoglobin + oxygen $\rightleftharpoons$ oxyhaemoglobin

1) **Haemoglobin** is a large, **globular protein** molecule made up of four polypeptide chains (see p.12 and 13).

2) Each chain has a **haem group** which contains **iron** and gives haemoglobin its **red** colour.

3) Haemoglobin has a **high affinity for oxygen** — each molecule carries **four oxygen molecules**.

'Affinity' for oxygen means willingness to combine with oxygen.

Partial Pressure *Measures* Concentration *of* Gases

The **partial pressure of oxygen** (pO_2) is a measure of **oxygen concentration**.
The **greater** the concentration of dissolved oxygen in cells, the **higher** the partial pressure.
Similarly, the **partial pressure of carbon dioxide** (pCO_2) is a measure of the concentration of carbon dioxide in a cell.

Oxygen **loads onto** haemoglobin to form oxyhaemoglobin where there's a **high pO_2**.
Oxyhaemoglobin **unloads** its oxygen where there has been a **decrease in pO_2**.

1) Oxygen enters blood capillaries at the **alveoli** in the **lungs**. Alveoli cells have a **high pO_2** so oxygen **loads onto** haemoglobin to form oxyhaemoglobin.

2) When our **cells respire**, they use up oxygen. This **lowers pO_2**, so red blood cells deliver oxyhaemoglobin to respiring tissues, where it unloads its oxygen.

3) The haemoglobin then returns to the lungs to pick up more oxygen.

Athletes train at **high altitude** because there's **low pO_2** in the air, which makes the body start producing **more red blood cells**. This means extra oxygen can be carried, so muscles respire more efficiently.

The extra red blood cells stay in the blood for a couple of weeks after returning to normal altitudes, giving the athlete an advantage. Sneaky.

Dissociation Curves *Show How* Affinity for Oxygen *Varies*

Dissociation curves show how the willingness of haemoglobin to combine with oxygen varies, depending on partial pressure of oxygen (pO_2).

100% saturation means every haemoglobin molecule is carrying the maximum of 4 molecules of oxygen.

0% saturation means none of the haemoglobin molecules are carrying any oxygen.

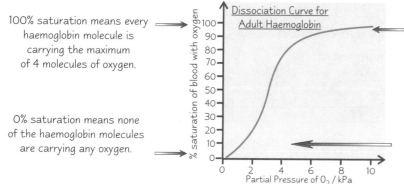

Where pO_2 is high (e.g. in the lungs), haemoglobin has a **high affinity** for oxygen (i.e. it will **readily combine** with oxygen), so it has a **high saturation** of oxygen.

Where pO_2 is low (e.g. in respiring tissues), haemoglobin has a **low affinity** for oxygen, which means it **releases oxygen** rather than combines with it. That's why it has a **low saturation** of oxygen.

The graph is 'S-shaped' because when haemoglobin (Hb) combines with the **first O_2 molecule**, it **alters the shape** of the Hb molecule in a way that makes it **easier** for other molecules to join too. But as the haemoglobin starts to become fully saturated, it becomes harder for more oxygen to join. As a result, the curve has a **steep** bit in the middle where it's really easy for oxygen molecules to join, and **shallow** bits at each end where it's harder for oxygen molecules to join.
When the curve is steep, a small change in pO_2 causes a big change in the amount of oxygen carried by the haemoglobin.

Haemoglobin and Oxygen Transport

Carbon Dioxide Levels Affect Oxygen Unloading

To complicate matters, haemoglobin gives up its oxygen **more readily** at **higher partial pressures of carbon dioxide** (pCO_2). It's a cunning way of getting more oxygen to cells during activity. When cells respire they produce carbon dioxide, which raises pCO_2, increasing the rate of oxygen unloading. The reason for this is linked to how CO_2 affects blood pH.

1) CO_2 from respiring tissues diffuses into red blood cells and is converted to **carbonic acid**.

2) The carbonic acid **dissociates** to give **hydrogen ions** and **hydrogencarbonate ions**.

3) If left alone, the hydrogen ions would increase the cell's acidity. To prevent this, oxyhaemoglobin **unloads** its oxygen so that haemoglobin can take up the hydrogen ions.

4) The **hydrogencarbonate ions** diffuse out of the red blood cells and are **transported in the plasma**.

5) When the blood reaches the **lungs** the low concentration of CO_2 causes the hydrogencarbonate and hydrogen ions to **recombine into CO_2**.

6) The CO_2 then diffuses into the **alveoli** and is breathed out.

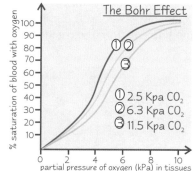

When carbon dioxide levels increase, the dissociation curve 'shifts' down, showing that more oxygen is released from the blood (because the lower the saturation of O_2 in blood, the more O_2 is being released). This is called the Bohr effect.

Fetal and Adult Haemoglobin have Different Affinities for Oxygen

In the **womb**, the fetus gets oxygen from its **mother's blood** across the placenta. This can only happen if its blood is **more likely** to absorb oxygen than its mother's blood. Because of this, **fetal haemoglobin** has a **higher affinity for oxygen** than adult haemoglobin. This means that fetal haemoglobin **always** has a slightly **higher saturation of oxygen** than adult haemoglobin, as you can see on the **graph**.

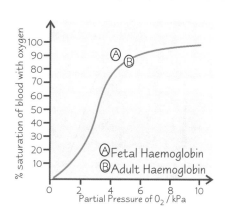

Practice Questions

Q1 How is HbO_8 formed?

Q2 What is pO_2?

Q3 What is carbon dioxide converted into in red blood cells?

Exam Questions

Q1 Explain why fetal haemoglobin is different from adult haemoglobin. [3 marks]

Q2 Explain the importance of the Bohr effect during a marathon race. [4 marks]

The Bore effect — it's happening right now...

Dissociation graphs can be a bit confusing — but basically, when tissues contain lots of oxygen (i.e. pO_2 is high), haemoglobin readily combines with the oxygen, so blood has a high saturation of oxygen (and vice versa when pO_2 is low). Simple. Also, make sure you get the lingo right, like 'partial pressure' and 'affinity' — hey, I'm hip, I'm groovy.

The Mammalian Heart

*Blood pumps **continually** round your body. It's happening right now as you read this rather long and dull sentence that I'm writing to keep you reading without stopping to try and highlight the **unceasingness** of blood flow. And relax.*

The **Heart** Consists of **Two Muscular Pumps**

The diagrams below show the **internal and external structure** of the heart.
The **right side** of the heart pumps **deoxygenated blood** to the **lungs**
and the **left side** pumps **oxygenated blood** to the **whole body**.

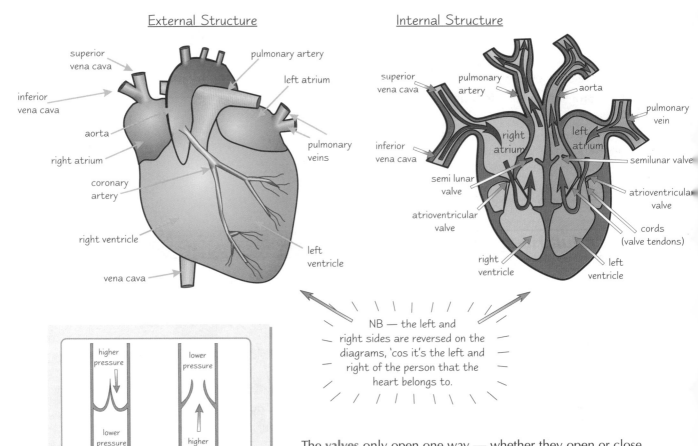

External Structure

Internal Structure

NB — the left and right sides are reversed on the diagrams, 'cos it's the left and right of the person that the heart belongs to.

The **valves** only open one way — whether they open or close depends on the relative pressure of the heart chambers.
If there's higher pressure behind a valve, it's forced open, but if pressure is higher above the valve it's forced shut.

Each Bit of the Heart is **Adapted** to do its Job **Effectively**

1) The **left ventricle** of the heart has thicker, more muscular walls than the **right ventricle**, because it needs to contract powerfully to pump blood all the way round the body. The right side only needs to get blood to the lungs, which are nearby.

2) The **ventricles** have thicker walls than the **atria**, because they have to push blood out of the heart whereas the atria just need to push blood a short distance into the ventricles.

3) The **atrioventricular** valves link the atria to the ventricles and stop blood getting back into the atria when the ventricles contract.

4) The **semilunar valves** stop blood flowing back into the heart after the ventricles contract.

5) The **cords** attach the atrioventricular valves to the ventricles to stop them being forced up into the atria when the ventricles contract.

Circulation

Mammals Have a Closed Double Circulation

Blood transports respiratory gases, products of digestion, metabolic wastes and hormones round the body in a **closed circulatory system**. It's called a **double** circulatory system because there are two routes:

1) **Systemic circulation**

   (heart ⟹ body ⟹ heart)

   Oxygenated blood travels to the body cells and **deoxygenated** blood returns to the heart.

2) **Pulmonary circulation**

   (heart ⟹ lungs ⟹ heart)

   Deoxygenated blood travels to the lungs and **oxygenated** blood returns to the heart.

Not all animals have double circulatory systems — some more simplistic animals like fish have single circulatory systems instead.

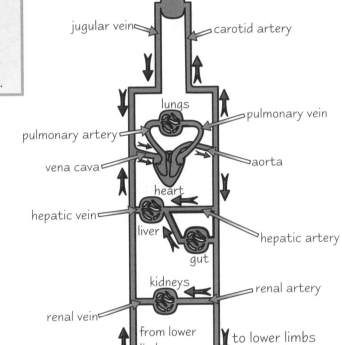

Mass Flow of Blood — Human Closed Circulatory System

This closed circulatory system uses the **heart** to pump blood round **blood vessels**.

Practice Questions

Q1　Which side of the heart carries oxygenated blood?

Q2　Why is the left ventricle more muscular than the right ventricle?

Q3　What is the purpose of heart valves?

Q4　What is the difference between the systemic and pulmonary circulatory systems?

Exam Question

Q1　Explain how valves stop blood going back the wrong way.　　　　　　　　　　　　　　　[6 marks]

Learn these pages off by heart...

Some of this will be familiar to you from GCSEs — so there's no excuse for not learning it really well.
The diagram of the heart can be confusing — it's like looking at a mirror image, so right is left and left is right.
(So in fact, when you look in the mirror you don't see what you actually look like — you see a reverse image — weird.)

Control of Heartbeat

You don't have to think consciously about making your heart beat — your body does it for you.
So you couldn't stop it beating even if for some strange reason you wanted to. Which is nice to know.

The **Cardiac Cycle** Pumps Blood Round the Body

The cardiac cycle is an on-going sequence of contraction and relaxation of the atria and ventricles that keeps blood continuously circulating round the body. The contraction and relaxation patterns alter the **volume** of the different heart chambers, which alters **pressure** inside the chambers. This causes **valves** to open and close, which directs the **blood flow** through the system.

There are 3 stages:

1 | Ventricles relax, atria contract

The **ventricles both relax**. The atria then contract, which decreases their volume. The resultant higher pressure in the atria causes the atrioventricular valves to open. This forces blood through the valves into the ventricles (**point 1** on the cardiac graph below).

2 | Ventricles contract, atria relax

The **atria relax** and the **ventricles then contract**. This means pressure is higher in the ventricles than the atria, which shuts off the atrioventricular valves to prevent backflow. Meanwhile, the high pressure opens the semilunar valves and blood is forced out into the pulmonary artery and aorta (**point 2** on the cardiac graph).

3 | Ventricles relax, atria relax

The **ventricles and the atria both relax**, which increases volume and lowers pressure in the heart chambers. The higher pressure in the pulmonary artery and aorta closes the semilunar valves to prevent backflow (**point 3** on the cardiac graph). Then the atria fill with blood again due to higher pressure in the vena cava and pulmonary vein and the cycle starts over again.

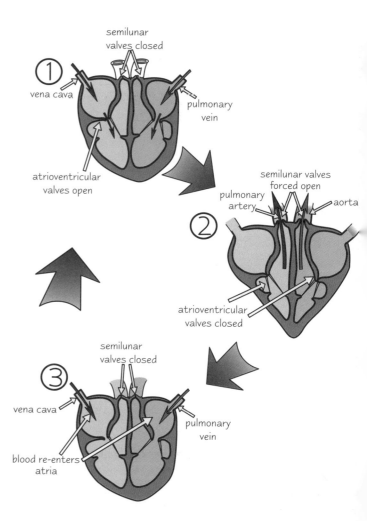

The **Cardiac Cycle** is Often Shown Using a **Graph**

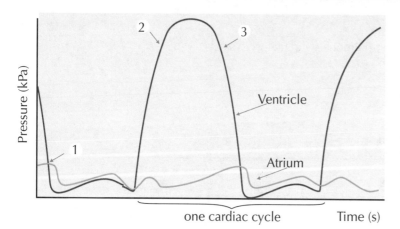

The graph shows the changes in **pressure** inside the heart chambers as they contract and relax. The lines represent the **ventricle** and **atrium**. Use the description of the cardiac cycle above to help you follow what's happening.

Make sure that you're familiar with this graph — it comes up pretty often in exams. It's better to learn it now than spending ages trying to get your head round it in the exam.

Control of Heartbeat

Cardiac Muscle Controls the Regular Beating of the Heart

Cardiac muscle is 'myogenic' — this means that, rather than receiving signals from nerves, it contracts and relaxes on its own. This pattern of contractions controls the regular heartbeat.

1) The process starts in the sino-atrial node (SAN) in the wall of the right atrium.

2) The SAN is a patch of modified muscle cells that acts like a pacemaker — it sets the rhythm of the heart beat by sending out regular electrical impulses to the atrial walls.

3) This causes the right and left atria to contract at the same time.

4) A band of non-conducting collagen tissue stops the electrical impulses from passing directly from the atria to the ventricles. This keeps the ventricles from contacting too early.

5) Instead, the atrio-ventricular node (AVN) picks up the impulses from the SAN. There is a slight delay before it reacts, so that the ventricles contract after the atria.

6) The AVN generates its own electrical impulse. This travels down into the finer fibrous tissue in the right and left ventricle walls, called Purkyne tissue.

7) The impulses mean both ventricles contract simultaneously, from the bottom up.

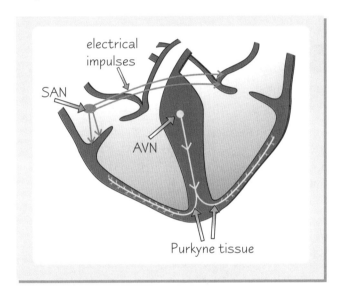

Sometimes the heart rhythm gets out of control. The wave of stimulation to heart muscle becomes chaotic. Different parts of the heart contract and relax at the same time. This is called fibrillation and can be fatal. Luckily, doctors can stop it by using defibrillating equipment, which sends an electric shock to the heart, returning it to its proper rhythm.

Practice Questions

Q1 Which valves in the heart are closed and which are open during the first stage of the cardiac cycle?

Q2 State whether the atria and ventricles relax or contract at each stage of the cardiac cycle.

Q3 During which stage of the cardiac cycle are the semilunar valves open?

Q4 What does "myogenic" mean?

Q5 What is the difference between the SAN and the AVN?

Q6 Why don't electrical impulses pass directly from the atria to the ventricles?

Exam Question

Q1 Describe the pressure changes which occur in the heart during contraction and relaxation. [3 marks]

The cardiac cycle — a stroke of myogenius, in my opinion...

Did you know that your heart can continue beating for a while after you're 'clinically dead' — spooky stuff. The reason is that it doesn't need any input from the brain and nervous system to operate. But, although the brain doesn't control heartbeat it does control the rate of the heartbeat. And that's quite enough about that.

Transpiration and Transport in Plants

Plants can't sing, juggle or tap-dance (as you will hopefully be aware). But they can exchange gases — how exciting. What makes it all the more thrilling is that they lose water vapour as they do it. Gripping stuff.

Transpiration is **Loss of Water** from a Plant's Surface

Transpiration is a side effect of gaseous exchange in plants. Plants need to exchange gases with the environment so they can **photosynthesise**, which uses carbon dioxide and water and produces oxygen. The carbon dioxide enters the plant through **stomata** — tiny pores in the surface of the leaf. Water **evaporates** from the moist cell walls and accumulates in the spaces between cells in the leaf. Then it diffuses out of the **stomata** when they open. This happens because there is a **diffusion gradient** — there's more water inside the leaf than in the air outside.

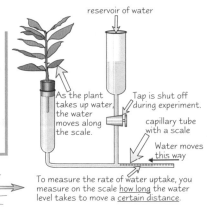

reservoir of water

As the plant takes up water, the water moves along the scale.

Tap is shut off during experiment.

capillary tube with a scale

Water moves <u>this way</u>

To measure the rate of water uptake, you measure on the scale <u>how long</u> the water level takes to move a <u>certain distance</u>.

<u>A potometer</u>

A <u>potometer</u> is a special piece of apparatus used to <u>measure transpiration</u>. It actually measures <u>water uptake</u> by a plant, but it's assumed that water uptake by the plant is <u>directly related</u> to water loss by the leaves. This lets you see how different factors affect the transpiration rate.

Four Main Factors Affect **Transpiration Rate**

The factors below affect transpiration rate. Temperature, humidity and wind alter the **diffusion gradient**, but **light** is a bit different:

So the rate of transpiration is fastest when it's light, warm, dry and windy.

1) **Temperature** — Diffusion involves the movement of molecules. Increasing the temperature speeds this movement up. So as temperature rises, so does transpiration rate.

2) **Humidity** — If the air around the plant is humid, the **diffusion gradient** between the leaf and the air is reduced. This slows transpiration down.

3) **Wind** — Lots of air movement blows away water molecules from around the stomata. This **increases** the diffusion gradient, which increases the rate of transpiration.

4) **Light** — Transpiration happens mainly when the stomata are open. In the dark the stomata usually close, so there's not much transpiration.

Two Types of **Tissue** Are Involved in **Transport in Plants**

Like most other multicellular organisms, plants need transport systems (see p.62). They have to be able to move substances like water, minerals and sugars from where they enter the plant (or where they're made in the plant) to where they're needed.

The **xylem** transports **water** and **mineral salts,** while the **phloem** transports dissolved substances, like **sugars.**   Xylem and phloem are found throughout the plant — they **transport materials** to all parts of the plant. Where they're found in each structure is connected to the **xylem**'s other function — **support**:

Plants also need <u>carbon dioxide</u> to make sugars by photosynthesis, but this enters the leaves (where photosynthesis happens) directly through the stomata.

1) In a **root**, which has to resist crushing as it pushes though the soil, the xylem is in the **centre**.

2) In **stems**, which need to resist bending, the xylem is **near the outside** to provide a sort of 'scaffolding'.

3) In a **leaf**, xylem and phloem make up a **network of veins**, which is needed for support because leaves are thin.

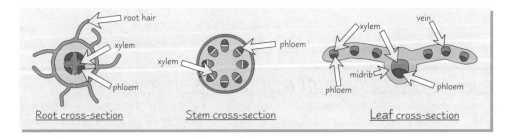

root hair

xylem

phloem

<u>Root cross-section</u>

xylem

phloem

<u>Stem cross-section</u>

xylem

vein

midrib

phloem

phloem

<u>Leaf cross-section</u>

Transpiration and Transport in Plants

Xylem Vessels Are Adapted For Transporting Water

Xylem tissue transports **water** and **mineral ions** up through plants. Xylem is a tissue made from several different cell types (see page 6), but you only need to worry about xylem vessels which are made up of the cells that actually transport the water and ions.

Xylem vessels are adapted for their function:

Xylem vessels are very long, **tube-like** structures formed from cells (**vessel elements**) joined end to end. There are **no end walls** on these cells, making an **uninterrupted tube** that allows water to pass through easily. The vessels are **dead**, containing **no cytoplasm**. Their walls are thickened with a woody substance called **lignin**, which helps with **support** and **stops** the xylem vessels **collapsing inwards**. The amount of lignin increases as the cell gets older. Substances get into and out of the vessels through small **pits** in the walls, where there is **no lignin**.

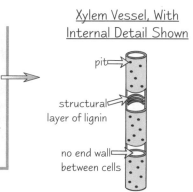

Xylem Vessel, With Internal Detail Shown

pit

structural layer of lignin

no end wall between cells

There are Two Main Cell Types in Phloem

Phloem tissue transports **solutes** (dissolved substances — mainly **sucrose**) round plants. Like xylem, **phloem** is formed from cells arranged in **tubes**, and these cells are modified for transport. But, unlike xylem, it's purely a **transport tissue** — it isn't used for support as well.

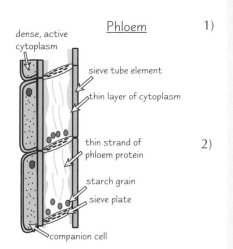

Phloem

dense, active cytoplasm

sieve tube element

thin layer of cytoplasm

thin strand of phloem protein

starch grain

sieve plate

companion cell

These are the two cell types you'll need to know about:

1) **Sieve tube elements** — These are **living cells** that **transport solutes** through the plant. They are joined end-to-end to form **sieve tubes**. The 'sieve' parts are the end walls, which have lots of holes in them. Unusually for living cells, sieve tube elements have **no nucleus** and only a **very thin layer of cytoplasm** without many organelles. The cytoplasm of adjacent cells is connected through the holes in the sieve plates.

2) **Companion cells** — The lack of a nucleus and other organelles in sieve tube elements means that they would have difficulty surviving on their own. So there is a companion cell for every sieve tube element. The companion cell has a very **dense and active cytoplasm**, and it seems to carry out the living functions for both itself and its sieve cell. Both are formed from a single cell during the development of the phloem.

Practice Questions

Q1 Define the term "transpiration".

Q2 What piece of apparatus is used to measure transpiration?

Q3 What are the two main cell types found in phloem tissue?

Exam Questions

Q1 Describe the distribution of the xylem and phloem in stems, roots and leaves.
Explain how this distribution is linked to the function of the xylem. [10 marks]

Q2 Describe factors that can alter the rate of transpiration in a plant, and explain the effect of each. [8 marks]

I'm sick of plants already — stupid boring green leafy mutter mutter birches mumble...

I'm sorry, it's just that (don't know if you've noticed) it's nearly the end of the book. And these plants are all that's left between me and freedom. Still, I suppose you need to know this stuff to do OK in your AS. So learn about xylem, learn about phloem, and don't forget transpiration either. Now if you'll excuse me, I'm off to kick a rhododendron for a while.

Water Transport

Water enters a plant through its roots and eventually, if it's not used, exits via the leaves. "Ah-ha," I hear you say, "— but how does it flow upwards, against gravity?" Well that, my friends, is a mystery that's about to be revealed.

Water Enters a Plant through its Root Hair Cells

Water has to get from the **soil**, across the **root** and into the **xylem**, which takes it up the plant. The bit of the root that absorbs water is covered in **root hairs**. This increases its surface area and speeds up water uptake. Once it's absorbed, the water has to get through two root tissues, the **cortex** and the **endodermis**, to reach the xylem.

> Water always moves from areas of **higher water potential** to areas of **lower water potential** — it goes down a **water potential gradient**. The **soil** around roots has a **high water potential** (i.e. there's lots of water there) and **leaves** have a **low water potential** (because water constantly **evaporates** from them). This creates a water potential gradient that keeps water moving through the plant in the right direction, **from roots to leaves**.

Look back to p. 23 for more on water potential gradients and the movement of water. But don't get your knickers in a twist about water potential — you won't be asked anything tricky about it in the exam, you just need to know what it is.

There are Three Routes Water can Take through the Root

Water can travel through the roots into the xylem by three different paths:

1) The **apoplast pathway** — goes through the **non-living** parts of the root — the **cell walls**. The walls are very absorbent and water can simply diffuse through them, as well as passing through the spaces between them.

2) The **symplast pathway** — goes through the **living** cytoplasm of the cells. The **cytoplasm** of neighbouring cells connects through **plasmodesmata** (small gaps in the cell walls).

3) The **vacuolar pathway** — water travels from the vacuole of one cell into the vacuole of the next by **osmosis**.

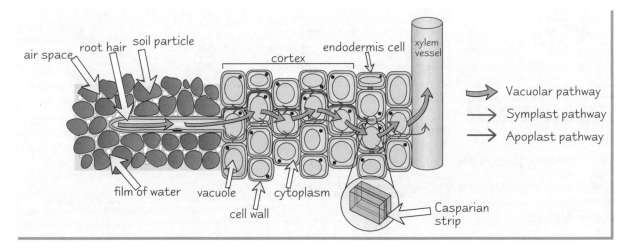

All three pathways are used, but the main one is the **apoplast pathway** because it provides the **least resistance**. When the water gets to the **endodermis** cells, though, the apoplast pathway is blocked by a **waxy strip** in the cell walls, called the **Casparian strip**, which the water can't penetrate. Now the water has to take one of the other pathways. This is useful, because it means the water has to go through a **cell membrane**. Cell membranes are able to control whether or not substances in the water get through (see p.21). Once past this barrier, the water moves into the **xylem**.

Water Transport

Water Moves Up a Plant Against the Force of Gravity

The **cohesion-tension theory** explains how water moves up plants from roots to leaves, against the force of gravity.

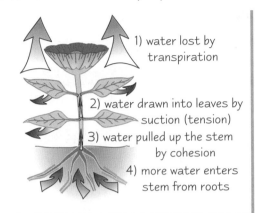

1) water lost by transpiration

2) water drawn into leaves by suction (tension)

3) water pulled up the stem by cohesion

4) more water enters stem from roots

1) Water evaporates from the leaves at the 'top' of the xylem (transpiration).

2) This creates a **suction** ('tension'), which pulls more water into the leaf.

3) Water molecules **stick together** ('cohesion' — see p.14), so when some are pulled into the leaf others follow.

4) This means the whole **column** of water in the xylem, from the leaves down to the roots, moves upwards.

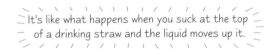

It's like what happens when you suck at the top of a drinking straw and the liquid moves up it.

Root pressure also helps move the water upwards. Water is transported into the xylem in the roots, which creates a pressure and tends to shove water already in the xylem further upwards. This pressure is weak, and couldn't move water to the top of bigger plants by itself. It helps though, especially in young, small plants where the leaves are still developing.

Explaining how Stan moves upwards against the force of gravity isn't quite as simple.

Practice Questions

Q1 How are roots adapted to absorb water?

Q2 Name the three pathways by which water travels across the root.

Q3 What is the Casparian strip and in which root tissue would you find it?

Q4 Name the theory that explains how water moves up a plant against the force of gravity.

Exam Questions

Q1 Explain the role of each of the following in the transport of ions in the root.

a) cell walls

b) the endodermis

c) plasmodesmata   [6 marks]

Q2 Explain why movement of water in the xylem stops if the leaves of a plant are removed.   [4 marks]

So many routes through the roots...

Lots of impressive biological words on this page, to amaze your friends and confound your enemies.
Go through the page again, and whenever you see a word like plasmodesmata, just stop and check you know
exactly what it means. (Personally I think they should just call them cell wall gaps, but nobody ever listens to me.)

Translocation

Translocation is the movement of **organic solutes** through a plant. It happens in the **phloem**.
Annoyingly, translocation sounds a lot like transpiration. Or is that just me? Don't confuse them anyway.

The Main Things *Translocated* are *Amino Acids* and *Sugars*

1) Sugars (mostly **sucrose**) are transported from the **leaves** (where they're made during photosynthesis) to **actively growing regions**, or to **storage sites**.

2) Amino acids are made in the **root tips** (where nitrogen is absorbed), and are carried to **growing areas** in the plant to **make proteins**.

The way that they're transported in the phloem isn't known exactly, but it is known that it's an active process, needing energy from respiration.

Translocation is the **movement of dissolved organic substances** (mainly sucrose) to **where they're needed** in the plant. Experiments have shown that translocation happens in the **phloem**.

Translocation moves substances from '**sources**' to '**sinks**'. The **source** of a substance is **where it's made** (so it's in **high concentration** there). The **sink** is the area where it's **used up** (so it's in **low concentration** there). For example, the source for sugars is the **leaves**, and the sinks are the other parts of the plant, especially the **food storage organs** and **growing points** in roots, stems and leaves.

Enzymes maintain a **concentration gradient** from the phloem to the sink by **modifying** the organic substances at the sink. For example, in **potatoes**, sucrose is converted to **starch** in the sink areas, so there's always a lower concentration of sucrose at the sink than inside the phloem. This makes sure a **constant supply** of new sucrose reaches the sink from the phloem.

The *Mass Flow Hypothesis* Best Explains *Phloem Transport*

It's still not certain exactly how the solutes are transported from source to sink by translocation. The best supported theory is the **mass flow hypothesis**.

The mass flow hypothesis:

1) Dissolved sugars from photosynthesis (e.g. sucrose) are **actively transported** into the **sieve tubes** of the phloem at the **leaves** (the **source**). This **lowers the water potential** inside the sieve tubes, so water enters the tubes by **osmosis**. This creates a **high pressure** inside the sieve tubes at the source end of the phloem.

2) At the **sink** end, **sugars leave** the phloem to be used up, which **increases the water potential** inside the sieve tubes, so water also leaves the tubes by **osmosis**. This **lowers the pressure** inside the sieve tubes.

3) The result is a **pressure gradient** from source end to sink end, which pushes sugars along the sieve tubes to where they're needed.

Mass Flow Hypothesis Can be *Demonstrated* in an *Experiment*

The idea behind the hypothesis can be shown in the experiment below.
In this model, **A** and **B** are two containers, each lined with a **selectively permeable membrane**.

1) **A** represents the **source** end and contains a **concentrated sugar solution**.

2) **B** represents the **sink** end, where **use** of the sugar **lowers** its concentration.

3) Water enters **A** by **osmosis**, causing the sugar solution to flow along the **top tube** — which represents the **phloem**.

4) **Hydrostatic pressure** increases in **B**, forcing water out and back through the **connecting tube** — which represents the **xylem**, because it just transports water.

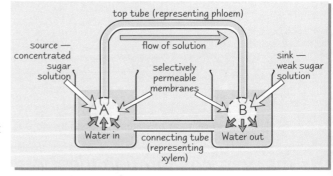

In the model, the **flow stops** when the sugar concentrations in the two containers **equal out**.
But in a plant this **wouldn't happen**, as sugar would constantly be produced by the source and used up by the sink.

Translocation

There is **Evidence** Both For and Against **Mass Flow**

Supporting evidence for this theory includes —

1)   There is a suitable **water potential gradient** between the leaves and the other parts of the plant.

2)   If the phloem is cut, sap oozes out. This shows that a **pressure gradient** does exist.

Objections to the theory are —

1)   Sugar travels to many different sinks, not just to the one with the **highest water potential**, as the model would suggest.

2)   The sieve plates would create a barrier to mass flow. A lot of pressure would be needed for the sugar solution to get through.

3)   Mass flow doesn't require living cells, yet the **phloem cells** are alive and very active.

Xerophytes are Plants that Live in **Dry Climates**

Xerophytes have adaptations which prevent them losing too much water by transpiration (see p. 72). Examples of xerophytic adaptations include:

1)   Stomata are sunk in **pits**, where water vapour is sheltered from wind.

2)   Leaves are **curled** with the stomata inside, again protecting them from wind.

3)   Layer of **'hairs'** on the epidermis traps moist air round the stomata, to reduce the diffusion gradient.

4)   **Reduced number of stomata**, so there are fewer places where water can escape.

5)   Thick, waxy, water-resistant **cuticle** on the epidermis to reduce water loss.

6)   **Swollen stem** to store water.

7)   **Roots** spread over a **wide area** just below the soil surface, to make the most of any rain.

8)   Whole leaf reduced to a **spike** (e.g. cactus). This reduces the surface area for water loss.

9)   **Take in CO_2 at night** so they can close their stomata in the day to reduce water loss, and use stored CO_2 for photosynthesis instead.

This book ain't over 'til the fat pig sings

Practice Questions

Ha, ha, ha, ha
Stayin' alive, stayin' alive

Q1   Explain the terms "source" and "sink" in connection with translocation.

Q2   State two pieces of evidence that support the mass flow hypothesis for translocation.

Q3   Name five adaptations seen in xerophytes that reduce the amount of water lost by transpiration.

Exam Questions

Q1   The mass flow hypothesis depends on there being a difference in pressure in the phloem sieve tubes between the source and the sink. Explain how sugars cause the pressure to increase at the source end, according to the mass flow hypothesis.   [4 marks]

Q2   The illustration shows a transverse section of a leaf of *Ammophila*, a xerophytic plant. State 4 ways, visible in the picture, that this leaf is adapted to dry conditions.   [4 marks]

Who cares whether the hypothesis is right — it's the last page...

_Blimey it's the end of the book. A sad, sad moment for biology students everywhere. But hey, at least the pig's happy.
He's still alive at the end of it all — he's survived genetic engineering, coronary heart disease and the nitrogen cycle, amongst
everything else. So I think we should all join in for a grand finale singsong. OK, maybe not — I think I need to go now._

Answers

Section 1 — Cell Structure
Page 3 — Electron and Light Microscopy

1   Maximum of 6 marks available.
    Advantages: Greater resolution / more detail *[1 mark]*.
    Greater magnification *[1 mark]*.
    Disadvantages: Electron microscopes can't be used to study living tissues *[1 mark]*.
    Natural colours can't be seen *[1 mark]*. They aren't portable *[1 mark]*. They are expensive *[1 mark]*.

Page 5 — Functions of Organelles

1   Maximum of 9 marks available.
    You could have written about any three of the five organelles given below.
    Mitochondria *[1 mark]* — Large numbers of mitochondria would indicate that the cell used a lot of energy *[1 mark]*, because mitochondria are the site of (aerobic) respiration, which releases energy *[1 mark]*.
    Chloroplasts *[1 mark]* — Large numbers of chloroplasts would be seen in cells that are involved in photosynthesis *[1 mark]* because the chloroplasts contain chlorophyll, which absorbs light for photosynthesis *[1 mark]*.
    Rough endoplasmic reticulum *[1 mark]* —
    You find a lot of RER in cells that produce a lot of protein *[1 mark]* because the RER transports protein made in the attached ribosomes *[1 mark]*
    Ribosomes *[1 mark]* —
    You find lots of ribosomes in cells that produce a lot of protein *[1 mark]* because ribosomes are the site where proteins are made *[1 mark]*
    Lysosomes *[1 mark]* — Found in cells that are old / destroy other cells *[1 mark]* because lysosomes contain digestive enzymes that can break down cells *[1 mark]*.
    There are 9 marks for this question and 3 organelles have to be mentioned — so it's logical that each organelle provides 3 marks. You get a mark for mentioning the correct organelles, so you need to give 2 pieces of relevant information for each one.

2 a) Maximum of 2 marks available.
    i) mitochondrion *[1 mark]*
    ii) Golgi apparatus *[1 mark]*
  b) Maximum of 2 marks available:
    The function of the mitochondrion is to be the site of (aerobic) respiration / provide energy *[1 mark]*.
    The function of the Golgi apparatus is to package materials made in the cell / to make lysosomes *[1 mark]*.
    The question doesn't ask you to give the reasons why you identified the organelles as you did, so don't waste time writing your reasons down.

Page 7 — Cell Organisation

1   Maximum of 6 marks available.
    1 mark for each example — e.g. ciliated epithelium, alveolar / squamous epithelium *[maximum of 2 marks]*.
    Up to 4 marks for explaining how their structure is linked to their function.
    E.g. Ciliated epithelium has hairs / cilia *[1 mark]* which move to waft substances along *[1 mark]*.
    It would be a good idea to give an example of what might be being moved (e.g. mucus, ova).
    Alveolar epidermis is thin *[1 mark]* so that respiratory gases can diffuse through it easily *[1 mark]*.
    The examples above are from the syllabus. You could also get the marks if you know of any other examples from your reading / lessons.

2   Maximum of 12 marks available.
    Up to 6 marks for correctly naming each leaf tissue.
    Up to 6 marks for explaining each tissue's adaptation:
    Lower epidermis *[1 mark]* — has stomata which allow carbon dioxide and oxygen in and out *[1 mark]*.

Spongy mesophyll *[1 mark]* — has air spaces which allow gases to circulate *[1 mark]*.
Palisade mesophyll *[1 mark]* — has many chloroplasts to absorb sunlight for photosynthesis *[1 mark]*.
Upper epidermis *[1 mark]* — waterproof to keep water in *[1 mark]*.
Xylem *[1 mark]* — delivers water to the leaf *[1 mark]*.
Phloem *[1 mark]* — transports sugars away from the leaf *[1 mark]*.

Section 2 — Biological Molecules
Page 9 — Carbohydrates

1   Maximum of 7 marks available.
    Glycosidic bonds are formed by condensation reactions *[1 mark]* and broken by hydrolysis reactions *[1 mark]*.
    When a glycosidic bond is formed in a condensation reaction, a hydrogen *[1 mark]* from one monosaccharide combines with a hydroxyl / OH group *[1 mark]* from the other to form a molecule of water *[1 mark]*.
    A hydrolysis reaction is the reverse of this *[1 mark]*, with a molecule of water being used up to split the monosaccharide molecules apart *[1 mark]*.
    The last 5 marks for this question could be obtained by a diagram showing the reaction, using structural formulae.

2   Maximum of 10 marks available.
    Glycogen is a chain of alpha glucose molecules *[1 mark]* whereas cellulose is a chain of beta glucose molecules *[1 mark]*.
    Glycogen's chain is compact and very branched *[1 mark]* whereas cellulose's chain is long, straight and unbranched *[1 mark]* and these chains are bonded together to form strong fibres *[1 mark]*.
    Glycogen's structure makes it a good food store in animals *[1 mark]*. The branches allow enzymes to access the glycosidic bonds to break the food store down quickly *[1 mark]*.
    Cellulose's structure makes it a good supporting structure in cell walls *[1 mark]*. The fibres provide strength *[1 mark]*.
    The function is helped by the fact that the cell doesn't have any enzymes that can break down beta bonds *[1 mark]*.
    In questions worth lots of marks make sure you include enough details. This question is worth 10 marks so you should include at least 10 relevant points to score full marks. Also, the question asks you to compare and contrast, so make sure you don't just describe glycogen and cellulose totally separately from each other. You need to highlight how they differ from each other, and what this means for their functions.

Page 11 — Lipids

1   Maximum of 10 marks available, from the points below.
    Up to 5 marks for correctly naming 5 functions of the 6 below.
    Up to 5 marks for explaining the relevant features of the 5 functions.
    Energy store *[1 mark]* — lipids contain a lot of energy per gram *[1 mark]*.
    Insulation *[1 mark]* — the lipid layer under the skin doesn't have an extensive blood supply so heat isn't lost from it *[1 mark]*.
    Buoyancy *[1 mark]* — the lipid layer is less dense than muscle and bone *[1 mark]*.
    Protection *[1 mark]* — the fat layer under the skin and around internal organs acts like a cushion to prevent damage from any blows *[1 mark]*.
    Waterproofing *[1 mark]* — lipids don't mix / dissolve in water, so water can't penetrate a lipid layer *[1 mark]*.
    Source of water for (desert) animals *[1 mark]* — water is released when lipids are respired *[1 mark]*.

2   Maximum of 8 marks available.
    A triglyceride consists of glycerol *[1 mark]* and three fatty acid molecules *[1 mark]*. A phospholipid has the same basic structure, but one of the fatty acids is replaced by a phosphate group *[1 mark]*.
    Triglycerides are hydrophobic / repel water *[1 mark]*. This is a property of the hydrocarbon chains that are part of the fatty acid molecules *[1 mark]*. The phosphate group in a phospholipid is hydrophilic / attracts water *[1 mark]*, because it's ionised *[1 mark]*. This means that the phospholipid has a hydrophilic 'head' and a hydrophobic 'tail' *[1 mark]*.

Answers

The words 'head' and 'tail' are not essential, as long as you have got across the idea that the molecule is partly hydrophobic and partly hydrophilic.

Page 13 — Proteins

1   Maximum of 9 marks available.
    Proteins are made from amino acids *[1 mark]*.
    The amino acids are joined together in a long (polypeptide) chain *[1 mark]*.
    The sequence of amino acids is the protein's primary structure *[1 mark]*.
    The amino acid chain / polypeptide coils in a certain way *[1 mark]*.
    The way it's coiled is the protein's secondary structure *[1 mark]*.
    The coiled chain is itself folded into a specific shape *[1 mark]*.
    This is the protein's tertiary structure *[1 mark]*.
    Different polypeptide chains can be joined together in the protein molecule *[1 mark]*.
    This is the quaternary structure of the protein *[1 mark]*.
    The question specifically states that you don't need to describe the chemical nature of the bonds in a protein. So, even if you name them, don't go into chemical details of how they're formed — no credit will be given.

2   Maximum of 6 marks available.
    Collagen is a fibrous protein *[1 mark]*.
    For this mark, mentioning that the molecule is fibrous is essential.
    It forms supportive tissues in the body *[1 mark]*.
    Collagen consists of three polypeptide chains *[1 mark]*.
    These chains form a triple helix *[1 mark]*.
    The chains are tightly coiled together *[1 mark]*.
    The tightly coiled chains provide strength to the structure *[1 mark]*.
    Minerals can bind to the collagen chain *[1 mark]*.
    This makes it more rigid *[1 mark]*.
    8 marks are listed, but the mark is given out of 6. This is common in longer exam questions. You would have to be a bit of a mind-reader to hit every mark the examiner thinks of, so to make it fair, there are more mark points than marks. You can only count a maximum of 6, though.

Page 15 — Water and Inorganic Ions

1   Maximum of 12 marks available, from any of the 15 mark points listed.
    Water molecules have two hydrogen atoms and one oxygen atom *[1 mark]*.
    The hydrogen and oxygen are joined by covalent bonds / sharing electrons *[1 mark]*.
    Water molecules are polar *[1 mark]*.
    Polarity leads to the formation of hydrogen bonds between water molecules *[1 mark]*.
    Water is a solvent *[1 mark]*.
    Water's polar nature allows water to dissolve polar solutes *[1 mark]*.
    Water transports substances *[1 mark]*.
    Substances are transported more easily when dissolved *[1 mark]*.
    Water has a high specific heat capacity *[1 mark]*.
    This is due to hydrogen bonds restricting movement *[1 mark]*.
    This means it's difficult to change the temperature of water *[1 mark]*.
    This allows cells to avoid sudden changes in temperature *[1 mark]*.
    Water has a high latent heat of evaporation *[1 mark]*.
    This is due to hydrogen bonding *[1 mark — but only awarded if hydrogen bonding hasn't already been mentioned]*.
    It means the evaporation of water has a considerable cooling effect *[1 mark]*.
    Be careful to stick to what the question asks for. The fact that ice floats on water and aquatic habitats have stable temperatures aren't things that happen "<u>in</u> living organisms".

2   Maximum of 7 marks available.
    Magnesium is a constituent of chlorophyll *[1 mark]*.
    If there's a lack of magnesium then chlorophyll can't be made, and the leaves will be pale or yellow *[1 mark]*.
    If there is less chlorophyll, there will be less photosynthesis *[1 mark]*.
    Low rates of photosynthesis will result in the plant growing less *[1 mark]*.
    Nitrogen is needed for the manufacture of proteins *[1 mark]*.

Proteins are need for growth *[1 mark]*.
A nitrogen shortage will lead to a shortage of protein, and the plant's growth will be stunted *[1 mark]*.
Although nitrogen's not mentioned in the table, you should have been able to deduce that nitrogen (N) is a component of nitrate. Cunning.

Section 3 — Enzymes
Page 17 — Action of Enzymes

1   Maximum of 8 marks available, from any of the 9 points below.
    If the solution is too cold, the enzyme will work very slowly *[1 mark]*.
    This is because, at low temperatures, the molecules move slowly and collisions are less likely between enzyme and substrate molecules *[1 mark]*.
    The marks above could also be obtained by giving the reverse argument — a higher temperature is best to use because the molecules will move fast enough to give a reasonable chance of collisions.
    If the temperature gets too high, the reaction will stop *[1 mark]*.
    This is because the enzyme is denatured *[1 mark]* — the active site changes and will no longer fit the substrate *[1 mark]*.
    Denaturation is caused by increased vibration breaking bonds in the enzyme *[1 mark]*.
    Enzymes have an optimum pH *[1 mark]*.
    pH values too far from the optimum cause denaturation *[1 mark]*.
    Explanation of denaturation here will get a mark only if it hasn't been explained earlier.
    Denaturation by pH is caused by disruption of ionic bonds, which destabilises the enzyme's tertiary structure *[1 mark]*.

Page 19 — Enzyme Activity

1   Maximum of 4 marks available.
    Put the potato and hydrogen peroxide solution in a test tube joined by a delivery tube to an upturned measuring cylinder in a bowl of water *[2 marks available for correctly naming all the apparatus; 1 mark if most of the apparatus is named]*.
    You then use the measuring cylinder to measure the amount of oxygen produced by the reaction *[1 mark]* over a specified time (e.g. one minute) *[1 mark]*.
    You could also draw an annotated diagram of the experiment to get all the marks. Other experiments are possible — for example, using a gas syringe.

2   Maximum of 4 marks available.
    Chemical X is an enzyme inhibitor *[1 mark]*
    Reason — it reduces an enzyme catalysed reaction *[1 mark]*
    The inhibitor is probably competitive *[1 mark]*
    Reason — increasing the concentration of the inhibitor makes it more effective, because if there are a lot of inhibitor molecules they're more likely to reach active sites before the substrate molecules and will block them *[1 mark]*.

Section 4 — Cell Membranes and Transport
Page 21 — The Cell Membrane Structure

1 a) Maximum of 1 mark available.
    In a triglyceride there are three fatty acids attached to a molecule of glycerol but in a phospholipid one of the fatty acids is replaced by a phosphate group *[1 mark]*.
  b) Maximum of 2 marks available.
    Phospholipids are arranged in a double layer / bilayer with fatty acid tails on the inside *[1 mark]*.
    Fatty acid tails are hydrophobic / non-polar so they prevent the passage of water soluble molecules through the cell membrane *[1 mark]*.
    Occasionally a question may ask you to show how a single layer of phospholipid molecules would arrange themselves on the surface of a container of water. You should draw the molecules with their hydrophilic phosphate heads in the water and their hydrophobic fatty acid tails sticking up into the air.

Answers

Page 23 — Transport Across the Cell Membrane

1   Maximum of 2 marks available.
    The ions are moving faster / have more kinetic energy **[1 mark]**, so
    more ions will diffuse through the membrane in a given time
    [1 mark].

Page 25 — Transport Across the Cell Membrane (continued)

1   Maximum of 10 marks available.
    The molecule might be too large **[1 mark]** or it might be charged
    [1 mark] or the concentration gradient across the membrane might
    be in the wrong direction **[1 mark]**.
    They might be transported by facilitated diffusion **[1 mark]**, which
    uses channel proteins for charged molecules and carrier proteins for
    molecules that are too large **[1 mark]**. It can only move molecules
    down a concentration gradient **[1 mark]**.
    Active transport also uses carrier proteins **[1 mark]**, but can work
    against a concentration gradient **[1 mark]**.
    Endocytosis is the process where substances are taken in by
    surrounding them with part of the cell membrane to form a vesicle
    [1 mark].
    Exocytosis is a similar process, but it happens in the opposite
    direction **[1 mark]**.

2 a) Maximum of 3 marks available.
    Root hair cell has a more negative / lower water potential than the
    soil water **[1 mark]** because it has a greater concentration of
    dissolved solutes **[1 mark]**.
    So water moves into the cell by osmosis from a higher water potential
    to a lower water potential **[1 mark]**.
  b) Maximum of 3 marks available.
    Shape increases the outer surface area of the cell membrane in
    contact with the soil **[1 mark]**.
    There are many carrier proteins in the membrane for active transport
    [1 mark]. There are many mitochondria in the cytoplasm for the
    production of the energy needed for active transport **[1 mark]**.
    When answering questions on adaptations, always link the physical
    feature to its function.

Section 5 — Genetic Control of Proteins
Page 27 — Structure of DNA and RNA / The Genetic Code

1   Maximum of 2 marks available.
    The long length and coiled nature of DNA molecules allows the
    storage of vast quantities of information **[1 mark]**.
    You only get the mark here if you've mentioned the length
    and the coiled nature.
    Good at replicating itself because of the two strands being
    paired **[1 mark]**.
    The question is worth two marks so you need to mention at
    least two things.

2   Maximum of 5 marks available.
    Nucleotides are joined by condensation reactions **[1 mark]**.
    This happens between the phosphate group and the sugar of the next
    nucleotide **[1 mark]**.
    The DNA strands join through hydrogen bonds **[1 mark]** between the
    base pairs **[1 mark]**.
    The final mark is given for at least one accurate diagram showing at
    least one of the above processes **[1 mark]**.
    As the question asks for a diagram make sure you do at least one, e.g.:

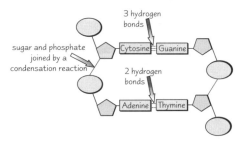

Page 29 — DNA Replication and Types of RNA

1   Maximum of 6 marks available.
    DNA strands uncoil and separate **[1 mark]**.
    Individual free DNA nucleotides pair up with their complementary
    bases on the template strand **[1 mark]**.
    DNA polymerase joins the individual nucleotides together
    [1 mark].
    Students often forget to mention this enzyme in their answers.
    Hydrogen bonds then form between the bases on each
    strand **[1 mark]**.
    Two identical DNA molecules are produced **[1 mark]**.
    Each of the new molecules contains a single strand from the original
    DNA molecule and a single new strand **[1 mark]**.

Page 31 — Protein Synthesis

1   Maximum of 2 marks available.
    A codon is a triplet of bases found on a mRNA molecule **[1 mark]**.
    An anticodon is a triplet of bases found on a tRNA molecule
    [1 mark].
    This question is only worth two marks so the examiner only
    expects a brief answer.

2   Maximum of 10 marks available.
    Transcription happens inside the nucleus and translation outside
    in the cytoplasm **[1 mark]**.
    A section of DNA is uncoiled by breaking hydrogen bonds **[1 mark]**.
    mRNA makes a copy of an uncoiled section of DNA **[1 mark]**.
    The mRNA travels outside the nucleus to a ribosome **[1 mark]**.
    The codons on the mRNA are paired with anticodons on a
    molecule of tRNA **[1 mark]**.
    The tRNA molecules are carrying amino acids **[1 mark]** which line up
    and are joined by peptide bonds with an enzyme **[1 mark]**.
    You must mention that an enzyme is involved to get this mark.
    Specific codons / base triplets code for specific amino acids **[1 mark]**.
    A polypeptide chain is formed **[1 mark]** — the primary structure
    of a protein **[1 mark]**.

Page 33 — Genetic Engineering

1   Maximum of 6 marks available.
    Haemophiliacs lack the healthy gene that codes for the protein factor
    VIII **[1 mark]**. The human gene that codes for this protein can be
    removed and inserted into the DNA of a bacterium to give
    recombinant DNA **[1 mark]**. An industrial fermenter is then used to
    culture the bacteria **[1 mark]**. They are given the ideal conditions
    needed for rapid growth and they reproduce quickly **[1 mark]**. They
    start producing the human protein, but as they can't use it, it builds
    up in the medium **[1 mark]**. It can then be extracted and processed
    for human use **[1 mark]**.

Section 6 — Nuclear Division
Page 35 — The Cell Cycle and Mitosis

1 a) Maximum of 6 marks available.
    A = Metaphase **[1 mark]**, because the chromosomes are lining up at
    the equator **[1 mark]**.
    B = Telophase **[1 mark]**, because the cytoplasm is dividing to form
    two new cells **[1 mark]**.
    C = Anaphase **[1 mark]**, because the centromeres have divided and
    the chromatids are moving to opposite poles **[1 mark]**.
    If you've learned the diagrams of what happens at each stage of mitosis,
    this should be a breeze. That's why it'd be a total disaster if you lost
    three marks for forgetting to give reasons for your answers. Always read
    the question properly and do exactly what it tells you to do.
  b) Maximum of 3 marks available:
    X = Chromatid **[1 mark]**.
    Y = Centromere **[1 mark]**.
    Z = Spindle fibre **[1 mark]**.

2 a) Maximum of 2 marks available.
    Interphase **[1 mark]**, during the S or Synthesis stage **[1 mark]**.

Answers

b) Maximum of 2 marks available:
Mitosis *[1 mark]*, during the prophase stage *[1 mark]*.
Sometimes you can pick up marks even if you're only half right. You might know spindle fibres are formed during mitosis, but you've forgotten which stage. It's worth mentioning mitosis anyway, and then having a guess. You won't lose marks if your guess is wrong, and even if it was you'd have picked up one mark just for saying mitosis. Better than leaving it blank.

Page 37 — Meiosis and Sexual Reproduction

1 a) Maximum of 3 marks available.
A = 46 *[1 mark]*.
B = 23 *[1 mark]*.
C = 23 *[1 mark]*.
b) Maximum of 2 marks available, from any of the points below.
Normal body cells have two copies of each chromosome, which they inherit from their parents *[1 mark]*.
Gametes have to have half the number of chromosomes so that when fertilisation takes place, the resulting embryo will have the correct diploid number *[1 mark]*.
If the gametes had a diploid number, the resulting offspring would have twice the number of chromosomes that it should have *[1 mark]*.

Section 7 — Energy and Ecosystems
Page 39 — Energy Transfer Through an Ecosystem

1 Maximum of 3 marks available.
100 units passes into the producers *[1 mark]*.
10 units pass into the primary consumers *[1 mark]*.
1 unit therefore passes into the secondary consumers *[1 mark]*.
Don't let daft little details like these 'arbitrary units' put you off. It just means that they're any old units, it doesn't matter what, and this actually makes the question easier for you.

Page 41 — The Nitrogen Cycle

1 Maximum of 10 marks available, from any of the following:
If a question asks you to 'give an account,' that means you have to describe what happens in words. But it never hurts to include a diagram, which will make things clearer for the examiner and could help jog your memory.
Mention the 4 processes that remove nitrogen from the atmosphere:
Nitrogen fixation by free living bacteria *[1 mark]*.
and those in root nodules (Rhizobium) *[1 mark]*.
Lightning *[1 mark]*.
Haber process *[1 mark]*.
Mention the 4 ways nitrogen enters and leaves the food chain:
Uptake of ammonium and nitrate ions by plants *[1 mark]*.
Consumers get nitrogen by feeding on other organisms *[1 mark]*.
Decomposers release ammonium ions *[1 mark]* when they feed on dead organisms *[1 mark]*.
2 marks for explaining nitrification:
Nitrosomonas convert ammonium ions into nitrite *[1 mark]* and Nitrobacter convert nitrite into nitrate *[1 mark]*.
2 marks for explaining denitrification: Conversion of nitrates to nitrogen gas *[1 mark]* by bacteria *[1 mark]*.
Make sure you know the difference between nitrification and nitrogen fixation, which often get confused. Nitrogen fixation is bacteria turning nitrogen gas into ammonium ions or amino acids. Nitrification is bacteria turning ammonium ions into nitrate.

Section 8 — Human Health and Disease
Page 43 — Health and Disease

1 Maximum of 4 marks available.
Up to 2 marks for giving advantages, and up to 2 marks for giving disadvantages.
Points you could include in your answer:
Advantages —
More accurate diagnoses of disease *[1 mark]*.
Development of better drugs, to target specific diseases *[1 mark]*.
Development of drugs with fewer side effects *[1 mark]*.

More information to base scientific research on *[1 mark]*.
Eradication of inherited diseases *[1 mark]*.
Less human suffering / early deaths *[1 mark]*.

Disadvantages —
Religious view of not altering 'God's work' *[1 mark]*.
Choice of 'designer' babies could lead to a change in gender percentage of population *[1 mark]*.
It's hard to draw the line between improving health and changing natural laws *[1 mark]*.
Any other suitable answer referring to ethical/moral issues.
Give a balanced answer — there are two marks available for advantages and two marks for disadvantages.

2 Maximum of 4 marks available. Any of the following points are acceptable.
Many areas of LEDCs have contaminated water supplies *[1 mark]*, poor medical care *[1 mark]*, poor education *[1 mark]*, poor housing *[1 mark]*, and malnutrition *[1 mark]*. Plus any other sensible answer.

Page 45 — A Balanced Diet and Essential Nutrients

1 Maximum of 3 marks available.
Equation EAR = BMR x PAL *[1 mark]*.
EAR = 7.2 x 1.4 = 10.08 (will accept 10.1) *[1 mark]* MJ per day *[1 mark]*.
(One mark for showing the equation / working, one mark for the correct answer (10.08), one mark for including units (MJ per day). You need to learn the equation for this — they won't give it to you in the exam. Always remember to show your working and put the correct units on at the end or you'll be throwing away easy marks. They've given you the units you need to use in the question (MJ per day), so make sure you use them in your answer.

2 Maximum of 4 marks available.
Essential amino acids can't be made in the body *[1 mark]*, so have to be included in food / diet *[1 mark]*.
Some can be converted into non-essential amino acids *[1 mark]* and all types of amino acids make proteins required in the body *[1 mark]*.

Page 47 — Malnutrition

1 a) Maximum of 2 marks available — 1 mark for correctly naming a disease, 1 mark for giving the correct cause(s):
EITHER: Marasmus *[1 mark]* — cause is PEM / lack of protein and energy *[1 mark]*.
OR: Kwashiorkor *[1 mark]* — cause is PEM / lack of protein and energy *[1 mark]*. Or any other sensible answer.
b) Maximum of 2 marks available — 1 mark for correctly naming a disease, 1 mark for giving the correct cause(s):
Obesity *[1 mark]* — cause is over-eating / high fat diet / not enough exercise / underactive thyroid gland *[1 mark]*.
Or any other sensible answer.

2 Maximum of 4 marks available.
Vitamin A is needed to make rhetonoic acid *[1 mark]*, which is needed for fighting pathogens *[1 mark]*.
It's also needed to make rhodopsin in the rod cells in the eyes *[1 mark]*, which is used to help us see in dim light *[1 mark]*.

Page 49 — Gaseous Exchange / Fitness

1 Maximum of 2 marks available.
With hypertension, blood pressure is too high *[1 mark]*.
This puts extra strain on the ventricular muscle and causes heart damage *[1 mark]*.

2 Maximum of 4 marks available.
Humans are large multicellular organisms *[1 mark]*.
The surface area : volume ratio is small in large organisms *[1 mark]*, which makes diffusion too slow *[1 mark]*.
Humans need specialised organs with a large enough surface area to keep all their cells supplied with enough oxygen and to remove CO_2 *[1 mark]*.

Answers

Page 51 — Energy Sources and Exercise

1   Maximum of 4 marks available.
Vigorous exercise means oxygen demand exceeds supply and the
body's skeletal muscles begin to respire anaerobically *[1 mark]*.
This produces lactic acid, which increases the toxicity in blood
[1 mark]. Removal of lactic acid by oxidisation to pyruvate requires
oxygen *[1 mark]* — when exercise stops this oxygen debt can be
repaid *[1 mark]*.

2   Maximum of 5 marks available, from any of the answers below:
Improved heart *[1 mark]* or lung *[1 mark]* function.
Resistance to disease *[1 mark]*.
Reduces excess weight *[1 mark]* and decreases cholesterol *[1 mark]*.
Improved posture *[1 mark]*.
Improved mental faculties *[1 mark]*.
Better muscle tone/less likely to strain muscles *[1 mark]*.
Loads of marks to choose from here — a lot of the time it's easy to drop
one, but with questions like this you need to make the most of it and grab
them all. Don't start putting down all the reasons you know, five will do
thanks, but if you couldn't think of five then you definitely need to go over
the page again. (If the way the question was phrased put you off, I forgive
you and you'll know for next time).

Page 53 — Effects of Smoking

1   Maximum of 4 marks available.
Passive smoking has been shown to be responsible for smoking
related deaths and poor health *[1 mark]*. Give examples of diseases
caused by tobacco smoke *[Up to 3 marks]*. Acceptable examples
include lung cancer, bronchitis, emphysema, atherosclerosis, coronary
heart disease, strokes, thrombosis.
For a question like this it would be a good idea to briefly say how inhaling
tobacco smoke causes these diseases, e.g. 'some of the toxic chemicals
from the tar in tobacco smoke are carcinogenic, so can cause lung cancer
when inhaled.' Note that this question is not asking you to argue for and
against banning smoking in public places.

2   Maximum of 4 marks available.
Chemicals in tobacco smoke destroy fibrous elastin tissue in alveoli
[1 mark]. This means the alveoli can't expand as air enters the lungs
[1 mark]. The smoke also increases the number of protein-eating
phagocyte cells, which digest the alveolar tissue *[1 mark]*.
This reduces the surface area for gas exchange *[1 mark]*.

Page 55 — Prevention and Cure

1   Maximum of 8 marks available.
A bypass operation uses sections of a vein taken from the leg *[1 mark]*.
These are attached to the heart *[1 mark]* so that the blood can flow
through them and bypass a blocked coronary artery *[1 mark]*. This
gives the heart muscle a better supply of oxygen and nutrients so it can
function better *[1 mark]*.
Transplant operations can give patients new hearts, but heart donors
are in short supply *[1 mark]*. If a patient's symptoms can be eased
with a bypass, this leaves donor hearts available for more urgent cases
[1 mark]. It's best not to do a transplant anyway unless it's absolutely
necessary, because the heart might be rejected by the patient's
immune system *[1 mark]*, which could kill them *[1 mark]*.
Bear in mind the examiner will give you marks for what you know, not
penalise you for mistakes. You won't lose marks if you include something
you're not sure about and it turns out to be wrong, but you'll gain them if
it's right.

Page 57 — Infectious Diseases

1   Maximum of 3 marks available.
Contact tracing allows quick identification of people with an
infectious disease who might have caught it from an infected person
or passed it onto them originally *[1 mark]*. Preventative measures can
be taken quickly *[1 mark]*, so the disease isn't transmitted *[1 mark]*.

2   Maximum of 4 marks available. Any of the following are acceptable.
Unprotected sex *[1 mark]* / mixing body fluids, like sweat and blood
[1 mark] / sharing needles for drug use *[1 mark]* / across the placenta
from mother to baby / in breast milk from mother the baby *[1 mark]*.
Any other sensible answer.

Page 59 — The Immune System

1   Maximum of 8 marks available.
The cellular response involves the T-lymphocytes *[1 mark]*.
T-lymphocytes recognise 'foreign' antigens on the surface of
pathogens, like bacteria, which cause disease *[1 mark]*.
Cytotoxic / killer T-cells stick to these antigens and secrete substances
that kill the pathogen *[1 mark]*.
Helper T-cells help plasma B-cells make antibodies and attract
macrophages which destroy the pathogen *[1 mark]*.
Memory T-cells have the ability to recognise an antigen in the future if
it re-enters the body *[1 mark]*. This enables a quicker response the
second time round *[1 mark]*.
Suppressor T-cells stop activity of killer T-cells and B-cells when
antigens have been destroyed *[1 mark]*.
The cellular response also triggers the humoral response *[1 mark]*.
This question has a lot of marks available. Don't worry if your answer is
organised differently to the one above (as long as you've included most of
the same points) but do make sure it has a sensible structure.

2   Maximum of 6 marks available.
Antibodies are produced by B-lymphocytes *[1 mark]* and are specific
to one antigen *[1 mark]*. They bind to the antigen *[1 mark]* forming
antigen-antibody complexes *[1 mark]* which can be disposed of by
macrophages *[1 mark]*. They can also rupture cells themselves, and
neutralise toxins *[1 mark]*.
Don't fall into the trap of thinking both parts of a question must always
be worth equal marks. Most of the marks for this question are for
describing the function of the antibodies.

Page 61 — Immunity and Vaccinations

1   Maximum of 7 marks available.
Vaccines use harmless antigens to stimulate an immune response
[1 mark]. **Must have a similar definition of a vaccine, and then any 3
of the 5 points given below, with the reason and it's explanation:**
However, some pathogens exist in many strains and keep changing by
mutation *[1 mark]*. This means that vaccines designed to protect
against earlier strains quickly become ineffective *[1 mark]*.
Some diseases (e.g. malaria) can live in animals as well as people
[1 mark]. A vaccination programme could eradicate such diseases
from the human population in an area, only for later generations to
become infected again by the animal vectors *[1 mark]*.
Similarly, there could be a section of the human population who
don't show any symptoms of a disease *[1 mark]*. It might appear to
have been eradicated, but if the vaccination programme is
discontinued, it could surface again in the population *[1 mark]*.
Stimulating the immune system through vaccination is not always
successful *[1 mark]* — e.g. some pathogens (such as cholera) invade
the gut, where the immune system doesn't work as well *[1 mark]*.
Vaccination programmes can be expensive and hard to organise
[1 mark], especially in Less Economically Developed Countries
[1 mark].

Section 9 — The Mammalian Transport System
Page 63 — Transport Systems / Blood Vessels

1   Maximum of 6 marks available.
Arteries *[1 mark]* — carry blood from the heart to the rest of the
body. They're muscular and have thick, elastic walls to cope with
high pressures caused by the heartbeat *[1 mark]*.
Capillaries *[1 mark]* — site where substances are exchanged between
the blood and the cells of the body. They are very small with walls
only one cell thick so that substances can diffuse efficiently between
capillaries and cells *[1 mark]*.
Veins *[1 mark]* — carry blood back to the heart. They're wide vessels
which contain valves to stop the blood flowing backwards *[1 mark]*.

Answers

Page 65 — Blood, Tissue Fluid and Lymph

1   Maximum of 3 marks available.
Phagocytes engulf pathogens / microorganisms and digest them
[1 mark].
They contain many lysosomes to aid digestion [1 mark].
They have elongated nuclei, and fluid cytoplasm so they can squeeze
through gaps to reach the site of infection [1 mark].
Exam questions relating to structure and function are a favourite.
Find a way (cartoons, poems, tables) to remember the name, structure
and function all together to avoid confusion.

Page 67 — Haemoglobin and Oxygen Transport

1   Maximum of 3 marks available.
The foetus relies on oxygen diffused from the mother's blood
[1 mark].
If the haemoglobin of the fetal and mother's blood had the same
affinity for oxygen they'd be competing for oxygen and diffusion
wouldn't happen [1 mark].
So fetal haemoglobin has a higher affinity for oxygen than its mother's
blood [1 mark].

2   Maximum of 4 marks available.
In a marathon, tissues are more active, so cells respire more quickly
[1 mark]. Oxygen is used up faster, creating a low pO_2 [1 mark].
Also, more CO_2 is released by cells, giving a higher pCO_2 [1 mark].
Both factors mean oxygen unloading from oxyhaemoglobin is
increased [1 mark].
Make sure you mention both pO_2 and pCO_2 in questions about
the Bohr effect.

Page 69 — The Mammalian Heart / Circulation

1   Maximum of 6 marks available.
The valves only open one way [1 mark].
Whether they open or close depends on the relative pressure of the
heart chambers [1 mark].
If the pressure is greater behind a valve (i.e. there's blood in the
chamber behind it) [1 mark], it's forced open, to let the blood travel
in the right direction [1 mark].
When the blood goes through the valve, the pressure is greater above
the valve [1 mark], which forces it shut, preventing blood from
flowing back into the chamber [1 mark].
Here you need to explain how valves function in relation to blood flow,
rather than just in relation to relative pressures.

Page 71 — Control of Heartbeat

1   Maximum of 3 marks available.
Pressure increases in atria when they contract and in ventricles when
they contract [1 mark].
Pressure decreases in atria when they relax and in the ventricles when
they relax [1 mark].
There is always more pressure on the left side of the heart due to
extra muscle tissue producing more force [1 mark].
This question doesn't ask you to describe the cardiac cycle — it
specifically asks you to describe the pressure changes during contraction
and relaxation. Mention both atria and ventricles in your answer.

Section 10 — Exchange and Transport in Plants
Page 73 — Transpiration and Transport in Plants

1   Maximum of 10 marks available:
Distribution can be explained in words or by diagrams, whichever
you find easier. In either case, these are the key points:
In the stem — Xylem and phloem towards the outside, with the
phloem outside the xylem [1 mark].
In the root — The xylem and phloem in the centre, with the phloem
outside the xylem [1 mark].
In the leaf — Main vein in the midrib, smaller veins in the rest of the
leaf, with xylem above the phloem [1 mark].
Function of xylem — to transport water and minerals [1 mark] and to
provide support [1 mark].

Stem is subjected to bending forces [1 mark].
This is best resisted by strengthening around the outside [1 mark].
Root is subjected to crushing forces [1 mark].
This is best resisted by strengthening in the centre [1 mark].
The leaf is thin and needs support throughout the tissue [1 mark].

2   Maximum of 8 marks available.
Up to 4 marks for correctly naming each factor. Up to 4 marks
for explaining the effect of each one.
Light [1 mark].
The stomata close in the dark, so transpiration cannot happen
[1 mark].
Temperature [1 mark].
Increasing temperature speeds up the movement of the
water molecules [1 mark].
Humidity [1 mark].
Increasing humidity reduces the diffusion gradient between the leaf
and the air, so transpiration slows down (or reverse argument for
decreasing humidity) [1 mark].
Air movement / wind [1 mark].
Dispersion of the water molecules in the air around the stomata
maintains the diffusion gradient [1 mark].

Page 75 — Water Transport

1   Maximum of 6 marks available:
a) Dissolved mineral ions pass through the cell walls en route to
   the xylem [1 mark].
   This is the apoplast pathway [1 mark].
b) Each endodermis cell has a waxy Casparian strip in its cell
   wall that water can't penetrate [1 mark].
   This blocks the apoplast pathway and allows selective
   absorption through the cell membrane [1 mark].
c) Plasmodesmata connect the cytoplasm of adjacent cells so
   dissolved ions can travel from cell to cell through them
   [1 mark].
   This is the symplast pathway [1 mark].

2   Maximum of 4 marks available:
Loss of water from the leaves due to transpiration pulls more
water in from xylem [1 mark].
There are cohesive forces between water molecules [1 mark].
These cause water to be pulled up the xylem [1 mark].
Removing leaves means no transpiration occurs, so no water is pulled
up the xylem [1 mark].
It's pretty obvious (because there are 4 marks to get) that it's not
enough just to say removing the leaves stops transpiration. You also
need to explain why transpiration is so important in moving water through
the xylem. It's always worth checking how many marks a question is
worth — this gives you a clue about how long your answer should be, and
how many details you need to include.

Page 77 — Translocation

1   Maximum of 4 marks available:
Sugars are actively transported into the sieve tubes at the
source end [1 mark].
This decreases the water potential of the sieve tubes [1 mark].
This causes water to flow in by osmosis [1 mark].
This means pressure is increased inside the sieve tubes at the source
end [1 mark].
I think this is a pretty nasty question. If you got it all right first time
you're probably a genius. If you didn't, you're probably not totally clear yet
about the pressure gradient idea. It's just that when a cell fills with fluid,
the molecules inside are under more and more pressure. If there's a high
concentration of sugar in a cell, this draws in water by osmosis, and so
increases the pressure inside the cell.

2   Maximum of 4 marks available.
Leaf is curled [1 mark].
Stomata are in pits [1 mark].
'Hairs' on the epidermis [1 mark].
Thick cuticle [1 mark].

Index

Index

Index